A Monograph of **Chalara** and Allied Genera

A Monograph of **Chalara** and Allied Genera

T. R. Nag Raj and Bryce Kendrick
Department of Biology
University of Waterloo
Waterloo, Ontario

COPYRIGHT © 1975
WILFRID LAURIER UNIVERSITY PRESS
WATERLOO, ONTARIO, CANADA
ISBN − 0-88920-027-0 − (cloth)

'*Narrow* species concepts with rather *broad* generic concepts offer a better approach to the understanding of a fungus flora than the multiplication of small monotypic genera'.
—A. H. Smith: Mycologia 60: 742-755, 1968.

We dedicate this book to that prince of a man and peerless mycologist, Dr. Martin B. Ellis, Chief Mycologist, Commonwealth Mycological Institute, England.

Preface

We began with the intention of monographing *Chalara* and very similar fungi (*Excioconidium,* etc.). We soon extended the scope of our study to encompass those dimorphic imperfect genera with *Chalara*-like phialides (*Thielaviopsis, Chalaropsis, Stilbochalara, Hughesiella*), then to cover two other genera with *Chalara*-like phialides but having characteristic ancillary sterile structures *(Chaetochalara, Sporoschisma).* Finally we considered several other genera with phialides having more or less cylindrical collarettes and deep-seated conidiogenous loci *(Bloxamia, Endosporostilbe, Ascoconidium, Sporendocladia)* or whose descriptions raised the suspicion that they had such attributes *(Endoconidium, Columnophora, Milowia).* To have extended the study further would have meant including phialidic genera which, although often having relatively deep collarettes, were otherwise not particularly morphologically comparable to *Chalara (Catenularia, Phialophora, Phialocephala, Sporoschismopsis).*

Our study, which began as an attempt to revise one genus, thus finished up by considering fifteen existing generic names (five reduced to synonymy, one to the status of *nomen dubium*) and adding one additional generic name (first published elsewhere).

Readers will find in the taxonomic part of this book a compilation of descriptions and illustrations of species of *Sporoschisma* and *Fusichalara.* Some may consider these redundant, especially since we have added nothing significantly new to what is already known about them: our purpose in including them is solely to bring together all taxonomic data relating to *Chalara* and other closely allied genera in a single source-book.

Acknowledgments

The task of bringing order into the taxonomy of a group beset with chaos could not have been accomplished without the courtesy and cooperation of the curators of the various international herbaria. We would particularly like to thank the following individuals for loaning collections in their keeping: Dr. W. Schultz-Motel, Botanisches Museum, Berlin-Dahlem, West Germany (B)*; Dr. C. R. Benjamin and Dr. M. L. Farr, The National Fungus Collections, Plant Industry Station, U. S. Dept. of Agriculture, Beltsville, Md., U.S.A. (BPI); Dr. E. Petit, Jardin Botanique de l'Etat, Bruxelles, Belgium (BR); Dr. A. Skovsted, Botanical Museum and Herbarium, Copenhagen, Denmark (C); Dr. J. T. Howell, Herbarium of California Academy of Sciences, San Francisco, Calif., U.S.A. (CAS); Dr. J. Parmelee, National Mycological Herbarium, Biosystematics Research Institute, Canada Agriculture, Ottawa, Ont., Canada (DAOM); Dr. A. Funk, Forest Research Laboratory, Victoria, B.C., Canada (DAVFP); Dr. I. Mackenzie-Lamb, Farlow Library and Herbarium of Cryptogamic Botany, Harvard University, Cambridge, Mass., U.S.A. (FH); Dr. M. B. Ellis, Commonwealth Mycological Institute, London, England (IMI); The Director, Instituto de Micologia, Universidade do Recife, Pernambuco, Brazil (URM); Dr. C. T. Rogerson, The New York Botanical Garden, Bronx Park, N.Y., U.S.A. (NY); The Director, Istituto Orto Botanico dell'Universita, Padova, Italy (PAD); The Director, Botanical Institute and Italian Cryptogamic Laboratory, The University, Pavia, Italy (PAV); Dr. R. Heim, Muséum National d'Histoire Naturelle, Paris (PC); The Director, Botanical Department of the National Museum, Prague, Czechoslovakia (PR); Dr. S. Ahlner, Botanical Department, Naturhistoriska Riksmuseet, Stockholm, Sweden (S); Dr. R. Santesson, Institute of Systematic Botany, University of Uppsala, Uppsala, Sweden (UPS); and Dr. C. G. Shaw, Washington State University, Pullman, Wash., U.S.A. (WSP). Dr. M. B. Ellis of the Commonwealth Mycological Institute and Dr. J. W. Carmichael of the University of Alberta Mold Herbarium, Edmonton, Canada, kindly supplied the cultures used in this study.

We are especially grateful to Dr. M. B. Ellis, without whose encouragement and generous loans of material this study would never have got off the ground, and also to Dr. S. J. Hughes, Biosystematics Research Institute, Ottawa, Canada for his helpful suggestions and criticisms, and for his collaboration with the treatment of *Fusichalara* and other New Zealand specimens. We thank Dr. Luella K. Weresub, Biosystematics Research Institute, Ottawa, Canada, for her always patient understanding and gentle guidance with our many problems of nomenclature. Dr. Kris Pirozynski, Biosystematics Research Institute, Ottawa, Canada, loaned specimens of *Chaetochalara aspera* and graciously permitted us to reproduce his drawing of an

---

*   The code names of the Herbaria are as published in the Index Herbariorum, Lanjouw and Stafleu, 1964 (Regnum Vegetabile Vol. 31).

associated discomycete.

Bryce Kendrick is deeply indebted to Miss Joan Dingley, Senior Mycologist at the Plant Diseases Division, D.S.I.R., Auckland, New Zealand (PDD), whose unfailing kindness and encyclopaedic knowledge made her the ideal companion on many fruitful forays during his sabbatical visit to New Zealand: trips from which many collections cited in this book were derived. Dr. Pim Sanderson, D.S.I.R., Lincoln, New Zealand, did much to make an extended collecting trip in South Island, New Zealand, both enjoyable and rewarding.

Ms. Sandie Cooper actively participated in these New Zealand forays, and her help in the subsequent examination and processing of the almost one thousand collections was truly invaluable.

S. B. P. Haag of the Classics and Romance Languages Department, University of Waterloo, very kindly corrected our numerous Latin diagnoses.

The National Research Council of Canada gave financial support to the research that enabled us to write this book. We are pleased to acknowledge this help. We thank the University of Waterloo for a subsidy that was largely instrumental in initiating the publication of this book and in bringing its price within the reach of many potential readers.

Ms. Sheena Curwood typed and retyped the manuscript with great patience and forbearance.

Contents

Introduction

The hyphomycete genus *Chalara*, as we interpret it here, includes such fungi as *Chalara quercina*, the conidial state of the oak-wilt fungus, *Ceratocystis fagacearum; Chalaropsis thielavioides,* known to be the causal agent of root rot of chinese elms, graft failure of walnuts, black mold disease of rose grafts, and storage rot of carrots; *Thielaviopsis paradoxa*, associated with pineapple disease of sugarcane, soft rot of pineapple, blackhead of bananas, stem bleeding and premature fruit fall in palms; *Thielaviopsis basicola*, a pathogen causing black root rot of tobacco and other economic hosts; and the unnamed conidial states of several species of *Ceratocystis* that are involved in decay of sweet potatoes and moldy rot of rubber *(Ceratocystis fimbriata)*, black rot of sugarcane *(Ceratocystis adiposa)*, and blue-gray discoloura-tion of pine and spruce *(Ceratocystis coerulescens* and *C. virescens)*.

Clearly the group is of some economic importance. We would like to point out that it also presents us with many prime examples of that fascinating phenomenon—and problem—known as 'pleomorphism'. The term does not simply cover those fungi with a sexual state and an asexual state. Many Fungi Imperfecti, whose sexual state is unknown, nevertheless produce more than one conidial state, and have thus over the years acquired either (1) more than one binomial or (2) one binomial deliberately and obligately based on the combined presence of two different conid-ial states. We are particularly concerned with the latter situation, which we deplore because it often forcibly separates the fungus concerned from other imperfect genera identical with one of its states, and thus introduces an illogical and confusing element to an already difficult situation.

The important generic names *Chalaropsis* and *Thielaviopsis*, which directly con-cern us in this work, are founded on just such a propinquity of states, in this case phialidic and chlamydosporic (Figure 1 B and C). So closely are these states juxta-posed that they often occur on the same hypha or conidiophore. The phialidic state, produced by these genera and also by the genus *Chalara* (Figure 1 A), very strongly suggests that the three genera are related. The other state, the chlamydo-sporic, present in *Thielaviopsis* and *Chalaropsis*, but lacking in *Chalara*, differen-tiates the first two genera from *Chalara*. The characteristic occurrence of these spores, in chains or singly, differentiates *Thielaviopsis* and *Chalaropsis*. Obviously the absence or loss of the second state would transfer either of the two genera into *Chalara*. El Ani (1958) has in fact reported such a case: a mutant of *Ceratocystis radicicola* whose conidial state differed from the norm in lacking 'aleuriospores' and thus had to be considered a *Chalara*. Significantly, we know of no case involving the loss of the phialidic state.

In dealing with such pleomorphic imperfects, Hughes (1953) suggested that the name should be based on the state that is most frequent and constant, most con-spicuous and most readily identifiable, with a proviso for referring to the secondary

spore types by the appropriate spore terminology. He indicated that for this to
work, it was necessary to establish the type of conidium development. Such an

1)   Diagrammatic interpretation of extant generic concepts of *Chalara, Chalaropsis* and
     Thielaviopsis.

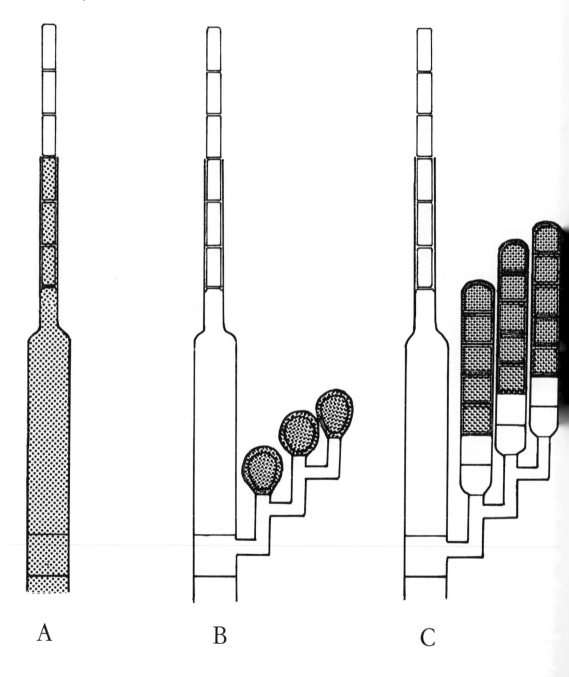

A   B   C

1

approach has obvious advantages and avoids much confusion. Hennebert (1968), redescribing the dimorphic fungus *Mammaria echinobotryoides*, considered the constant occurrence of typical 'aleuriospores' as diagnostic of the fungus, but stated that its phialidic state could also be a new form species of *Phialophora*. In the case of *Thielaviopsis* and *Chalaropsis*, the true nature of the second state characterized by thick-walled, pigmented propagules does not seem to be well understood. Differing terms—macroconidia, A-conidia, exospores, aleuriospores, chlamydospores—have been employed to describe them. Information about their ontogeny is basic to determining their taxonomic value—whether we should use them to segregate taxa at the generic or at the infra-generic level.

Barron (1968) was one of the few who recognized the arbitrary nature of the distinctions between the three genera. He subscribed to the view that if a form-genus needs to be circumscribed around one state only, then *Thielaviopsis* and *Chalaropsis* must be considered as synonyms of *Chalara*, since all three have a phialidic state which is essentially the same. We have concluded that the wider occurrence of a highly characteristic phialidic state provides us with a more rational generic concept, but it does not necessarily exclude the provision of a generic name or names for the chlamydosporic states. What we do deplore is the provision of binomials for each and every conidial state of a pleomorphic imperfect.

This book deals with several such pleomorphic genera and particularly the monomorphic genus *Chalara* to which we have assigned their species. We hope that our solution of these problems, and our reasons for it, will persuade other mycologists to apply a similar rationale when treating other pleomorphic imperfect genera.

In addition to the *Chalara* states of *Ceratocystis adiposa, C. autographa, C. fimbriata, C. moniliformis, C. radicicola* and *Cryptendoxyla hypophloia*, sixty-one species of *Chalara* are recognized, including twenty-six new species and three new combinations. Five genera are reduced to synonymy with *Chalara*, and six others with affinities to *Chalara* are discussed. Four taxa are excluded from *Chalara*, and eleven others considered as *nomina dubia*.

Materials and Methods

This study is based on an examination of microscopic preparations made from about 315 type and other specimens, and a number of living cultures. Most of the specimens mentioned in this work were kindly loaned by the curators of the herbaria listed in the acknowledgments.

Some of the material received from the various herbaria contained prepared slides in which the material had been stained with cotton blue in lactophenol. Where necessary, supplementary semi-permanent slide preparations were made in lactophenol, mostly without staining, to facilitate observation under phase contrast illumination.

Live cultures of *Thielaviopsis basicola* and unnamed cultures of *Chalara* were received from UAMH (University of Alberta Mold Herbarium, Edmonton). IMI supplied cultures of 'endoconidial' *Ceratocystis* species, whose identity was confirmed by reference to material authentic for the names.

To obtain reasonable comparisons with some of the herbarium material, cultures of *Chalara* spp., *T. basicola*, and *Ceratocystis* spp., were grown on thin autoclaved strips of *Pinus strobus* wood in vials; slide preparations were made from such cultures after adequate growth and sporulation had occurred.

The colour nomenclature given in the descriptions is not drawn from colour standards such as Ridgway because of the growing consensus that such standards are not reliable (Illman and Hamly 1948). In this study the differentiation of closely related shades is not important, though relative depth of pigmentation sometimes is.

As far as possible we endeavoured to obtain adequate samples for observation, but sometimes the poor condition of the material in old specimens prevented this.

When one space of the ocular micrometer scale equals 12.4 μ/ 3.12 μ/ 1.28 μ, the 'probable' error of any measurement is 3.1 μ/ 0.78 μ/ 0.32 μ, respectively. For most of the measurements reported the standard deviation is between 10% and 20% of the general mean.

Cultures were grown and maintained on potato-dextrose agar and malt extract agar. Ontogeny of thick-walled propagules in *Thielaviopsis basicola*, *T. paradoxa* and *Ceratocystis radicicola* was studied in living cultures by time-lapse photomicrography. Initially the slide culture chamber of Cole and Kendrick (1968) was used, but the petri dish technique of Cole, Nag Raj and Kendrick (1969) was subsequently found to be more satisfactory.

History, Morphology and Conidium Ontogeny

The morphological features used in a classification must be relatively stable under environmental pressure. In the following pages we present a brief history of *Chalara* and 'related' genera, listing and evaluating stable morphological features, and describing conidium ontogeny as a basis for delimiting taxa. Finally, we indicate our disposition of each generic name before passing on to the next.

Chalara (Figures 1-2, 12-36, 43A)
At this point we are discussing the concept of the genus *Chalara sensu stricto* (i.e., as recognized before this study). Corda (1838) employed the name *Chalara* for a fungus on a conifer at Breznia, Czechoslovakia, giving it the rank of a subgenus in *Torula* (Pers.) Link, with *Torula (Chalara) fusidioides* Cda. as the type species. His description was as follows: "Stroma grumulosum, effusum, floccis simplicissimis, hyphopodio lageniformi uniloculari suffultis et e sporiis cylindricis non septatis compositis." Its inclusion in Torulaceae Cda., characterized by him as "fungi e sporiis concatenatis continuis compositi, hyphopodio vel stromate spuriis vel nullis suffulti" and his illustration of the fungus published in Icones Fungorum, reflect his original inaccurate belief that the conidia of *Chalara* arise by fragmentation of the fertile hyphae. He considered *Chalara* adequately distinguishable from other members of *Torula* by means of "die Unterlage und die Hyphopodienform." Corda (1842) gave *Chalara* the rank of a section* of *Torula*, but Rabenhorst (1844) soon raised it to generic rank with a diagnosis—"Unterlage verbreitet, krümlig; Flocken einfach, auf einem flaschenformigen Träger, später in Walzenformige, einfache Sporen zerfallend," which endorsed Corda's ideas.

Saccardo (1880) threw some light into this murky corner of mycology when he introduced a totally new and much more accurate generic concept, keying *Chalara* out with the character "conidia catenulata *ex interiore hypharum exsilientia* ."* Höhnel (1902), in a regressive step, divided the genus into three genera or subgenera: *Euchalara* with exogenous, acrogenous and catenulate conidia; *Endochalara* with endogenous conidia that emerge in chains from the apex of the fertile hyphae; and *Synchalara* with the short fertile hyphae densely connate on a thin subiculum and the conidia as in *Endochalara*. Fortunately, this misconception did not remain

---

*   Nag Raj and Hughes (1974) have pointed out that Reichenbach (1841) listed 'Chalara Corda' and 'Torula P.' as infrageneric divisions of 'Torula P.' and that if Pfeiffer's (1873) interpretation of Reichenbach's usage of the two names as sections of 'Torula P.' was correct, then Reichenbach's usage antedates that of Corda (1842) of *Chalara* as a section of *Torula*; but Reichenbach's intended infra-generic category for *Chalara* is not clear.

*   Italics are ours.

long unchallenged. Lindau (1907) rightly objected to Höhnel's specious subdivision of the genus, since the exogenous conidia of *Euchalara* would clearly exclude it from *Chalara*, and rejected the names *Euchalara, Endochalara* and *Synchalara*. He recognized the origin of the conidia within the conidiogenous cells as the characteristic feature of the genus and wrote: "the mechanism by which the conidia are formed in the interior of the conidiogenous cells ('Büchsen') is not precisely known, but is probably similar to endogenous spore formation in many other fungi. Probably the upper cells are hollow and the conidia are formed in the interior by acrogenous septation, emerging in the form of chains from the aperture above."

There has been no subsequent attempt to reassess the genus which, at the time this study began, comprised 46 names, including two at the infraspecific level, three *nomina nuda* and two of uncertain status. Nag Raj and Kendrick (1971) and Nag Raj and Hughes (1974) have added nine more species to the genus.

In most species of *Chalara,* the vegetative mycelium is immersed in the natural substrate. In a few species, however, it is superficial and composed of a loose or compact network of hyphae, which appear hyaline, subhyaline or pale brown, and septate with the septa either closely spaced or relatively distant. The walls mostly appear smooth, but in a few species (e.g., *C. cylindrica, C. emodensis*) they are verrucose to finely echinulate. The thickness of the wall varies with the species. Some species (e.g., *C. selaginellae*) form aggregations of short, broad, dark brown, thick-walled cells arranged in thin, flat layers of prosenchymatous or rarely pseudoparenchymatous nature which are either superficial on the epidermis of the host or, in a few instances, subepidermal.

Very few species of *Chalara* have been studied in culture. Henry (1944) reported loosely knit, irregularly shaped, tan to black sclerotial masses in 3-month-old cultures of *Chalara quercina.*

The *conidiophores* are *phialophores.* They arise as lateral outgrowths of individual or aggregated vegetative hyphae or from short, broad, thick-walled cells of pseudoparenchymatous layers. In conidial states of a few *Ceratocystis* species (e.g., *Chalara ungeri, C. quercina*) they are hardly distinguishable from the vegetative hyphae until the liberation of conidia begins, while in others they are morphologically distinct. When borne directly on the vegetative hyphae, they are usually solitary and scattered or sometimes gregarious, but when they originate from aggregations of hyphae or cells of prosenchymatous layers (Figure 2E) they are arranged in loose or compact fascicles. In some species (e.g., *C. fusidioides, C. ampullula*), the conidiophores are simple conidiogenous cells, actually phialides (see below), arising directly on the superficial vegetative hyphae (Figure 2A), though in *C. fusidioides* an occasional phialide is borne on a simple, cylindrical structure of 1-3 cells (Figure 30F). When the conidiophore is composed of a phialide borne on a supporting structure, the latter may be a simple, short, cylindrical or obconical cell (Figure 2B, C), or it may be a simple cylindrical column of few to many cells (Figure 2D). The basal cell of the conidiophore in a few species (e.g., *C. cylindrosperma*) is often slightly or moderately swollen (Figure 2D). The conidiophore may or may not be constricted at the septa. It usually appears smooth, but is verrucose or finely echinulate in a few species (e.g., *C. cylindrica, C. emodensis, C. bohemica*). In many instances the conidiophore is much lighter in colour at the apex than elsewhere. Proliferation of the conidiophores does not appear to occur in many species. When present, it is sympodial as in *C. bohemica* (Figure 29C). Occasionally, regeneration of a conidiophore may occur several times through the broken ends (percurrent proliferation), to give rise to a fresh crop of conidia, as in *C. insignis.*

Diagrammatic representation of the morphology of *Chalara* species. A-E Phialophore types;
F-I. Phialide types; J-L. Types of collarette; M-T. Types of conidium.

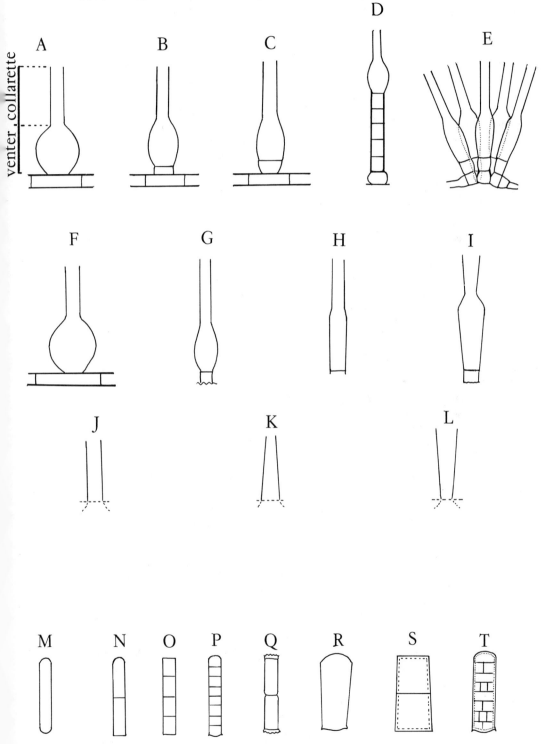

The characteristic *conidiogenous* cell of *Chalara* is a *phialide** of peculiar morphology. The basic features of this cell are a more or less expanded lower portion, or *venter* (see Figure 2A), and a narrower, more or less tubular, open ended *collarette* (see Figure 2A). Within *Chalara*, four morphological variations on this theme may be distinguished: ampulliform or lageniform (Figure 2F), obclavate (Figure 2G), subcylindrical (Figure 2H) and urceolate (Figure 2I). In the ampulliform or lageniform type, the venter is globose, subglobose or ellipsoidal, being widest at its median section, and much narrower above and below. In the obclavate and subcylindrical types, the venter is less marked, being short- or long-cylindrical, and of more or less even width along its length. In the urceolate type, the venter is obconical, with a narrow base corresponding to the width of the terminal cell of the conidiophore, and wider above. The collarette is morphologically distinct from the venter in most *Chalara* species, but in a few, like *C. ungeri* (Figure 32B) and *C. quercina* (Figure 32A), it is not clearly differentiated. When distinct, the collarette is cylindrical (Figure 2J) as in *C. fusidioides*, conical (Figure 2K) as in the *Chalara* state of *Ceratocystis virescens*, or obconical (Figure 2L) as in *C. breviclavata*. Depending on the morphological type of phialide involved, the transition from venter to collarette may be abrupt (often marked by a perceptible constriction as in *C. ampullula* [Figure 30D], *C. insignis* [Figure 17A]), gradual, or barely perceptible (e.g., *Chalara* state of *Ceratocystis virescens*). The wall of the phialide may be of uniform thickness except for the distal part of the collarette where it becomes attenuated (e.g., *Chalara ampullula* [Figure 30D], *C. fusidioides* [Figure 30F]) or it may be thickest at the base of the phialide, gradually becoming thinner towards the distal end of the collarette. In most species, the wall of the venter is smooth, but in *C. bohemica* (Figure 29C), *C. cylindrica* (Figure 29B) and *C. emodensis* (Figure 19B) it is often finely echinulate. The wall of the collarette is also smooth in most species; in *C. inflatipes* (Figure 13B) and *C. insignis* (Figure 17A), however, a verrucose appearance of the distal part of the collarette is due to the presence of dense, but minute, transverse striae. The collarette is darker than the venter in *C. nigricollis* (Figure 20B), *C. inflatipes* (Figure 13B) and *C. insignis* (Figure 17A), but in some others it is lighter than the venter and the rest of the conidiophore.

Cylindrical *enteroblastic-phialidic* conidia are characteristic of most species of *Chalara*. In *C. breviclavata*, the conidia are shortly clavate (Figure 2R) and in *C. cladii* they are subcylindrical (Figure 2S). The ends of the conidia are mostly blunt or truncate (Figure 2O), but in a few species the apex is rounded (Figure 2N), and in a few others both ends are rounded (Figure 2M). In species producing conidia with a truncate base, a minute but distinct marginal frill is often present (Figure 2P, R), e.g., *C. acuaria*, *C. brachyspora*, *C. rivulorum*, *C. selaginellae*, *C. breviclavata*. In species like *C. inflatipes*, *C. insignis*, *C. dictyoseptata* and *C. pulchra*, this frill is prominent. The conidia of *C. rubi* (Figure 18B) are unique in having a fringe of wall material extending for some microns from the base or from both ends. While the conidia are unicellular (ameroconidia) in most species (Figure 2M, R), they are consistently 1-septate (didymoconidia) in a few (Figure 2N, Q, S), and variably septate in a few others (Figure 2O, P) as in the 0-4-septate phragmoconidia of *C. pteridina*, and even muriformly septate (dictyoconidia) in *C. dictyoseptata* (Figure 2T). In conidia with more than one cell, the cells are usually of equal length, but are unequal in *C. cladii* and *C. inaequalis* (Figures 18A and 20A). Constriction of the conidium wall at the septum has been seen only in *C. rubi* (Figure 18B). The conidium wall is usually thin and smooth in most species. In species like *C. pulchra*,

---

*   A PHIALIDE is defined as a *determinate* conidiogenous cell which produces, from a *fixed conidiogenous locus*, a basipetal succession of *enteroblastic* conidia whose walls arise de novo. For a more extensive definition of phialide, and for explanations of the words in italics above, see Kendrick 1971, chapter 16.

C. inflatipes, C. cladii and *C. dictyoseptata*, however, the conidium walls are comparatively thicker; smooth in the first two and verrucose in the last two. Subhyaline to pale brown conidia are found in *C. cladii, C. selaginellae,* and *C. dictyoseptata,* though the conidia of most other species are hyaline. Conidium length ranges from a minimum of 2 μ in *C. brevispora* to a maximum of 66 μ in *C. bicolor.* In species with small, unicellular conidia, the conidia are usually extruded in long but easily disrupted chains. In species with larger, several-celled conidia, the conidium chains are perceptibly shorter, or even absent, as in *C. inflatipes* and *C. dictyoseptata.*

Unger (1847) appears to have been the first investigator to recognize that the conidia are endogenous in *Chalara.* He was looking at *C. ungeri,* though he had mistakenly identified the collection as *Graphium penicillioides* Cda. In 1891 Brefeld described conidium formation in a fungus, subsequently named *Chalara brefeldi* by Lindau (1907), as follows: "Die Konidien treten wie der Büchse heraus; man sieht erst dicht hinter ihrer Mündung eine auscheinend fertige Sporen liegen, welche dann durch eine zweite hinter ihr entstehende herausgeschoben wird." Hughes (1953) noted the similarity of conidium development in *Sporoschisma* and *Chalara insignis.* He wrote: "The stalked phialides in *S. mirabile* Berk., . . . consist of a bulbous base leading above to a 1-3-septate stalk which then enlarges into a slightly swollen part which then narrows again into a more or less cylindrical tube closed at the rounded apex. Up to three phialospores are laid down within the cylindrical tube from the apex backwards; finally the thinner-walled apical cap of the phialide is torn off presumably by pressure from within. The uppermost phialospore has a rounded apex and is usually shorter and has fewer septa than the others and when pushed out from below is usually found with the torn off apex of the phialide capping its distal end. A similar type of development can be observed also in *S. saccardoi* Mason and Hughes, *S. juvenile* Boud. and in *Chalara insignis* (Sacc., Rouss., and Bomm.) comb. nov."

We tried to study the conidium ontogeny in the single viable culture of *Chalara* sp. (UAMH 1548) available, by time-lapse photomicrography. Unfortunately, the fungus either produced only vegetative hyphae in the slide culture chamber of Cole and Kendrick (1968) or, in the agar culture chamber of Cole, Nag Raj and Kendrick (1969), produced vegetative hyphae so profusely as to mask the developing conidiophores and render observations or photomicrography difficult. Our observations are, therefore, confined to herbarium material showing conidiophores in different stages of development and are illustrated diagrammatically in Figure 8I (page 46). (Fortunately, other fungi which we now place in *Chalara* were less recalcitrant—see section on *Thielaviopsis*).

In the early stages of development, the phialide or conidiophore initial appears as a lateral branch of a vegetative hypha. In those species with multiseptate conidiophores, the branch continues to elongate and the terminal cell is differentiated as a phialide after the requisite number of septa have been laid down. The phialide initial is separated from the basal supporting structure by a septum, and continues to elongate with or without increase in width, then gradually or abruptly narrows and finally elongates into a cylindrical, subcylindrical or obconical apical part. At least one well-differentiated conidium is seen beneath the apex and a few more conidia may be differentiated below it. The basal wall of the youngest conidium is in contact with the apical wall of the conidium being formed at the conidiogenous locus below. Presumably because the new conidia developing below it exert pressure on the first conidium, the apical wall of the phialide breaks by a circumscissile split and the first conidium escapes. New enteroblastic conidia are then extruded in basipetal succession, forming at the conidiogenous locus which is found at the point where the venter and collarette meet. The shape of the conidia is largely determined by the shape of the collarette. In species with small, unicellular conidia, several differentiated conidia may be seen within the collarette, while in species with large, multi-septate conidia, only one or a few can be accommodated there. In species with septate conidia, the septa are often formed well before the conidia are released. How the marginal frill at the base of the conidia, or the fringes of wall material seen on conidia of *Chalara rubi,* develop is not clear.

We consider the species, whose morphology we have discussed above, to form the backbone of our newly expanded generic concept of *Chalara.*

Thielaviopsis (Figures 1, 3, 4, 39-42)

Went (1893) established *Thielaviopsis* for a fungus on sugarcane in Java, and gave the following generic description: "Hyphae steriles repentes, subhyalinae, fertiles simplices, septatae. Conidia dimorpha, majora catenulata, ovata, fusca; minora cylindracea, hyalina, ex interiore hypharum catenulatim generata et ex apice exsilientia." He described *Thielaviopsis ethacetica* Went as the single species of the genus. De Seynes (1886) had already described and illustrated *Sporoschisma paradoxum* de Seynes on pineapple. Saccardo (1892), considering it to be a *Chalara,* proposed the combination *Chalara paradoxa* (de Seynes) Sacc. Höhnel (1904) observed a fungus identical with *S. paradoxum* growing on mildewed endosperm of coconut and considered it to be the same as *T. ethacetica.* At Höhnel's request, Went examined it and found that it was indeed *T. ethacetica.* Because de Seynes' epithet had priority, Höhnel (1904) proposed the combination *Thielaviopsis paradoxa* (de Seynes) Höhnel. Delacroix (1893) described *Endoconidium fragrans* Delacr., from a collection of pineapple in Paris, but Höhnel (1909) brought this name into synonymy with *T. paradoxa.* Mitchell (1937) named an Australian isolate from diseased banana, *T. paradoxa* (de Seynes) Höhn. var. *musarum* Mitchell (*nom. nud.*), which, according to him, differed from var. *paradoxa* in the size of its 'exospores' and 'endospores' on natural substrates, its pathogenicity, and its temperature relationships. Riedl (1962), describing a new species of *Ceratocystis, C. musarum* Riedl, from bananas in storage in Vienna, considered the conidial state under the name *T. musarum* Riedl and brought *T. paradoxa* var. *musarum* into synonymy with it.

De Seynes (1886) described *S. paradoxum* as growing in the parenchyma of fruit of *Ananas,* but did not indicate the origin of the specimen or the herbarium in which it was deposited. Our attempts to trace de Seynes' material have not been successful. But an analysis of his account and illustrations of the fungus provides adequate diagnostic information. The following features are apparent (freely translated from de Seynes):

1) The vegetative mycelium is composed of narrow, hyaline, septate hyphae.
2) The conidiophores are erect, narrow at the base, abruptly enlarging above it and then attenuated less abruptly towards the apex, of a uniform smoky tint, with 3 or 4 septa in their enlarged basal part. A linear series of juxtaposed, hyaline, cylindrical conidia occurs in the upper part of these conidiophores.
3) Conidiophores formed later are of the same smoky tint or lighter and give rise to pale brown cylindrical conidia of the same dimensions as the preceding, or larger and more rounded, and wider than the conidiophores.
4) On the same mycelium, and occasionally side by side with the conidiophores described above, short, erect, cylindrical, hyaline conidiophores arise and produce deep greenish brown conidia in short chains from their apex.
5) The conidia are unicellular. The first formed conidia are colourless and appear white in mass, while those formed later are brown and appear black in mass.
6) On a single conidiophore, one may occasionally see a chain of conidia in which the terminal conidium is oval and brown with a flattened base; some of the

remaining conidia are hyaline and cylindrical, and others rounded and brown like the terminal conidium.

De Seynes called the coloured conidia 'macroconidia,' since they are wider than the conidiophores, and considered their development endogenous, but less obviously so than that of the hyaline cylindrical conidia. A comparison with our Figures 3, 4, and 41 will show excellent agreement. We believe there can be little doubt that we have correctly divined the nature of de Seynes' fungus. The following comparative observations on the morphological features of *Thielaviopsis paradoxa* are based on several collections disposed under the name.

Vegetative hyphae are subhyaline, septate, and branch irregularly. Occasional hyphal elements become dark brown, with walls up to $1\,\mu$ thick (Figure 41) and may be regarded as chlamydospores. The conidiophores arise as erect, lateral branches of the hyphae, from which they are not easily distinguishable until they are mature. The phialophores are simple, cylindrical to subcylindrical, septate, hyaline or subhyaline, terminating in subcylindrical, obclavate or lageniform phialides with long collarettes. Transition from venter to collarette is gradual or barely perceptible. The walls of the phialides are smooth or minutely verrucose, and concolorous with the rest of the phialophores. The conidiogenous locus occurs at or above the median part of the phialide. New phialides or phialophores arise irregularly as lateral branches from just below the septa in the basal part of the conidiophore. Lax, erect, short coremia composed of dense aggregations of conidiophores often occur sparsely in cultures of the fungus. In older cultures, the conidiophores that arise directly from the hyphae, or as separate lateral branches from the basal cells of existing conidiophores, are generally shorter, terminating in a relatively shorter cylindrical or subcylindrical phialide which has a very short collarette and the conidiogenous locus close to the apex of the phialide (Figure 4M).

Two morphological kinds of conidia are recognizable. The first kind are enteroblastic-phialidic conidia, cylindrical or doliiform with rounded, blunt or truncate ends; unicellular, hyaline, subhyaline or pale brown, extruded in long chains from the apex of the collarette. An occasional conidium that has become brown and slightly thick-walled is seen at or just within the apex of the collarette. The second kind of conidia are ovoid, pyriform or ellipsoidal, unicellular, brown or reddish brown, with a vertical or oblique germ slit, borne singly or in short chains on short, septate or non-septate conidiophores. New lateral branches, themselves producing similar solitary, terminal, thick-walled conidia, may arise just below the septa of the conidiophores. Höhnel (1904) observed, as did Went (1893), transitional stages between the two conidial forms, often in the same chain. In cultures, a conidiophore that has borne a single, terminal, thick-walled conidium may proliferate laterally to form a new phialide bearing enteroblastic, short-cylindrical conidia. Eventually, the phialides and hyphae collapse and the reddish brown, thick-walled conidia give a dark powdery appearance to the senescent cultures.

Cole and Kendrick (1969) investigated the development of the phialoconidia in a culture of *Thielaviopsis paradoxa* (conidial state of *Ceratocystis paradoxa*) in its early phase of growth and listed the following basic characters of conidium formation:

"1) During the early stage of differentiation, growth of the outer wall and conversion of the protoplast proceed simultaneously. One to several conidia differentiate basipetally within the cylindrical sporogenous cell. Apical growth ceases, and the center of activity continues to move downward until the protoplasmic conversion is exactly balanced by protoplasmic growth, that is, when the 'endogenous meristem' assumes its final position.

2) The conidia escape by rupturing the outer wall at the apex of the sporogenous cell.

3) The collarette may be so long that it encloses several physiologically independent conidia.

4) Many conidia are formed in a basipetal succession from the fixed conidiogenous meristem. This meristem is located deep within the body of the sporogenous cell and is not marked by any constriction or other morphological differentiation.

5) The first formed conidium is morphologically distinct from all others.

6) The conidiophore does not elongate after its apex has ruptured."

We have made our own time-lapse studies of *T. paradoxa* (see Figures 3, 4), and have thus been able to provide information on the subsequent behaviour of the fungus.

Figures 4N-Q show further changes that occur in cultures of this age. In Figure 4N, a phialide and a large number of cylindrical phialoconidia have become dark brown and thick-walled. Figure 4O shows a chain of doliiform phialoconidia that have become dark brown. In Figure 4P, a chain of cylindrical, hyaline, enteroblastic conidia formed by a phialide have begun to change colour, and the individual phialoconidia show a slight increase in width (Figure 4Q).

Figure 3A shows three conidiophores, each with 1 or 2 basal septa. The conidiogenous cell terminating each of these conidiophores is more or less cylindrical and shorter than the normal phialides. From each of these conidiogenous cells two physiologically independent conidia have been delimited; the terminal conidium is broadly ovoid and the second conidium is doliiform. A third conidium is being formed enteroblastically at the apex. The apex of the conidiogenous cell is indicated by an arrow, and its basal septum by an arrowhead, as reference points. After 1200 minutes many conidia have been delimited basipetally and form long chains (Figure 3F). If the distance between the two reference points is measured in each of the 6 photomicrographs, it will be found that there has been no change in the length of the conidiogenous cell, which is clearly a phialide. In Figure 3B, the terminal conidium in each of the three chains has begun to darken, and in Figure 3F all conidia but the youngest have become dark brown and thick-walled. Figures 4A-F present another time-lapse sequence covering growth over a period of 2310 minutes. Figure 4A shows a sparsely septate, narrow, cylindrical conidiophore. The septum separating the conidiogenous cell from the rest of the conidiophore is indicated by an arrowhead, and the conidiogenous locus by an arrow. The first conidium initial is being formed at the swollen apex of the conidiogenous cell. After 155 minutes (Figure 4B), a second conidium has been differentiated beneath the first, which has been displaced upwards. Figure 4C shows five mature conidia, two of which (the first two formed) have become dark brown, while a third has begun to darken. These conidia no longer remain in a chain. The length of the conidiogenous cell has not changed. Four hours later (Figure 4D), a sixth conidium has been fully differentiated and three of the six conidia have become dark brown. The conidiogenous cell has subsequently proliferated sympodially. The point of initiation of the new growth is shown by an arrowhead, and the new conidiogenous locus by an arrow; initiation of a conidium at the newly proliferated apex is seen in the form of a bulbous swelling. After a further 1210 minutes three more conidia have been produced by the new conidiogenous locus and all have become dark brown and thick-walled (Figure 4F). Figure 4M clearly shows the phialidic nature of the conidiogenous cell. The phialide is subcylindrical or cylindrical with a very short, flaring collarette (arrow) inside which the bulging apical wall of the conidium initial is seen; the conidiogenous locus is slightly beneath it. The phialide has produced seven physiologically independent conidia, four of which have begun to

) *Chalara paradoxa.* 35mm time-lapse sequence of enteroblastic-phialidic chlamydospore formation at a late phase of culture development. Arrowhead = reference point; arrow = apex of phialide. A-F at 0, 50, 150, 200, 290 and 1,200 minutes respectively.

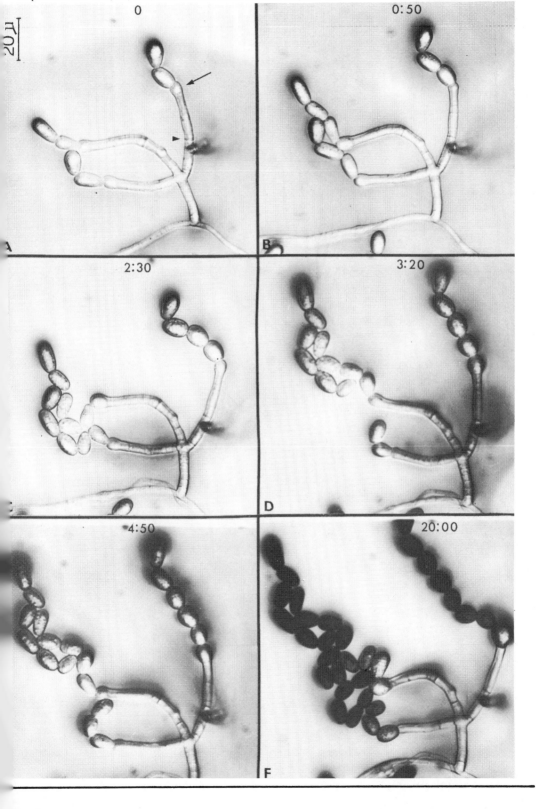

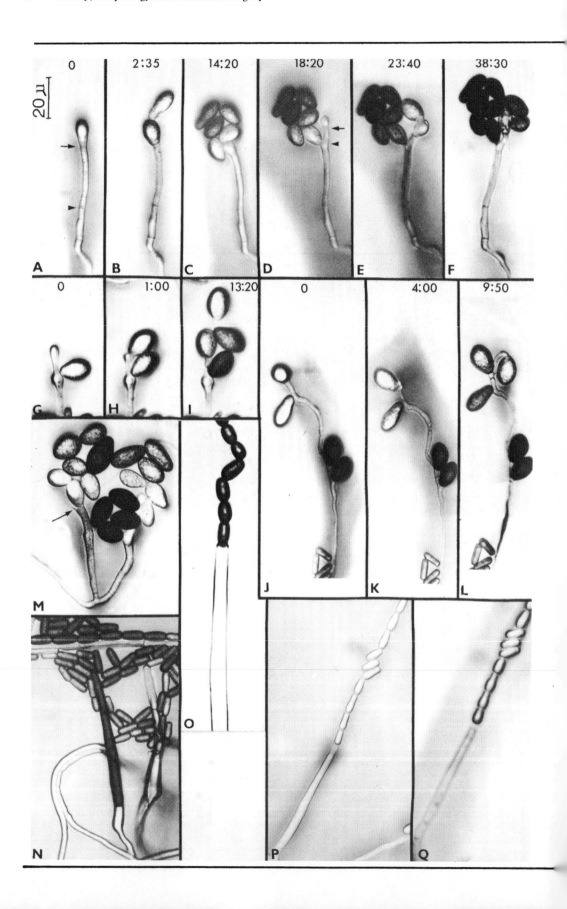

darken and to acquire thick walls.

A different conidium development, one foreshadowed in Figure 4D, can be seen in Figures 4G-I and J-L. In Figure 4G the conidiogenous cell has produced a mature conidium from its swollen apex and has proliferated percurrently before producing a new conidium 60 minutes later (Figure 4H). A slight bulge indicates that it is beginning to proliferate again. In Figure 4I, 740 minutes later, the proliferating conidiogenous cell has produced four conidia, three of which show increasing degrees of pigmentation. Figures 4J-L show repeated sympodial proliferation of the conidiogenous cell producing a single, ovoid conidium at each level. The conidia shown in these two sequences are morphologically indistinguishable from the phialidic kind shown in Figure 3 and 4A-F, and usually occur when the cultures are 4-5 days old and small air pockets have appeared between the edge of the cover slip and the drying, thin layer of agar. If the inside of the cover of the petri dish is lined with moist filter paper and the culture kept covered for 4-5 hours, a conidiophore of the kind shown in Figure 4K reverts to the production of phialoconidia as in Figure 4D, probably as a result of the high humidity.

In reviewing these facts, the following stages in conidium development can be discerned:

1) Initially the fungus produces conidiophores that are hardly distinguishable from the vegetative hyphae. One to several physiologically independent, enteroblastic-phialidic conidia are differentiated basipetally within the cylindrical portion of the conidiogenous cell after it ceases extension growth. The first conidium escapes by rupturing the outer wall at the apex of the conidiogenous cell, and the conidiogenous locus then becomes fixed at or slightly above the median part of the conidiogenous cell, which can now be regarded as a phialide, and goes on to produce a basipetal succession of enteroblastic conidia. The first-formed conidium is morphologically distinct from the others, which are cylindrical. These conidia may subsequently become dark brown and thick-walled.

2) Conidiophores formed when the cultures are older than 24 hours,* are shorter, less robust, and sparsely septate in the basal part. The phialide is far shorter, and subcylindrical or cylindrical. A single conidium is differentiated at the swollen apex of the phialide. Apical growth ceases, and the first conidium escapes by rupturing the phialide apex. The collarette is very short and obconical, enclosing a single conidium initial. The first-formed conidium is broadly ovoid and somewhat different from subsequent conidia, which are doliiform or ovoid and may or may not remain in a chain. The phialide does not necessarily elongate once its apex is ruptured, but in fact often proliferates percurrently or sympodially. Vegetative proliferation may be followed by further conidiogenesis. The conidia become dark brown and thick-walled with age (Figure 3). The rate of conidium

---

*   Times given here are those prevailing in the agar culture chamber, but may be slightly different in normal petri dish cultures depending upon the depth or composition of the medium, temperature, humidity, etc.

Chalara paradoxa. A-L. 35 mm time-lapse sequence of chlamydospore formation at a very late phase of culture development. Arrowhead = reference point; arrows = conidiogenous locus. A-F at 0, 155, 860, 1,420 and 2,310 minutes; G-I at 0, 60, and 1,800 minutes; J-L at 0, 240 and 590 minutes. M. Phialide with arrow indicating the flaring collarette and the bulging apical wall of the conidium initial; N. Thick-walled and darkly pigmented phialide and phialoconidia; O. Chain of doliiform, darkly pigmented, thick-walled phialoconidia; P-Q show increase in width, darkening and thickening of the walls of phialoconidia.

production is now declining.

3) In cultures more than 48 hours old, new conidiophores are at first extremely short, and the conidiogenous cell is no longer a phialide. It is narrow, cylindrical, and passes through the same early phases of development described in the preceding sections for the phialide. A conidium initial is delimited at its swollen apex, and apical growth ceases. However, the wall of the conidiogenous cell does not subsequently rupture, but itself becomes the conidium wall; the conidium is separated from the rest of the conidiogenous cell by a basal septum. Conidium development thus becomes holoblastic. Renewed activity of the conidiogenous locus beneath the terminal conidium results in sympodial proliferations of the conidiogenous cell, each culminating in the formation of a solitary, terminal holoblastic conidium at a higher level. Depending upon conditions, the conidiophore may proliferate repeatedly and the arrangement of the conidia appears to be acropetal. These conidia are morphologically identical to the ovoid conidia seen in Figures 3 and 4A-F. The rate of conidium production is extremely low at this stage. The three stages of conidial development are depicted diagrammatically in Figure 8, I-III.

In general, the fungus exhibits an extreme degree of ontogenetic plasticity, initially producing enteroblastic-phialidic conidia, then shifting in older cultures to holoblastic conidium formation. Regardless of whether they are the cylindrical, doliiform or ovoid phialoconidia, or the ovoid holoblastic conidia, all become dark brown and thick-walled, and function as chlamydospores, as do some of the phialides (Figure 4N) and hyphal elements.

There is no evidence to support Went's conclusion (1893) that the brown, thick-walled conidia arise as arthroconidia.

Berkeley and Broome (1850) published *Torula basicola* Berk. & Br. for a fungus found on *Pisum* sp. and *Nemophila auriculata* in England, with a brief description accompanied by an illustration of the fungus. Zopf (1876), observing an ascomycete associated with *Torula basicola*, gave it the name *Thielavia basicola* Zopf, believing that the two fungi were related. Following Zopf's publication, *Torula basicola* was often referred to in error as *Thielavia basicola* (Berk. & Br.) Zopf, though Zopf had not proposed such a combination. Ferraris (1910) transferred *Torula basicola* to *Thielaviopsis*, forming the combination *Thielaviopsis basicola* (Berk. & Br.) Ferr. [sub *Thielaviopsis basicola* (Berk.) Ferr. (sic)], but listed it as the conidial state of *Thielavia basicola* with *Helminthosporium fragile* Sorok. as its synonym. McCormick (1925) showed that there was no genetic connection between *Thielaviopsis basicola* and *Thielavia basicola*, despite their close physical association. Ferraris's transfer of *Torula basicola* to *Thielaviopsis* received no recognition until the publication of McCormick's paper.

Petri (1903) added *Thielaviopsis podocarpi* Petri to the genus, from a collection on roots of *Podocarpus* sp. in Italy, and considered his fungus distinct from *T. basicola* in the branching of the fertile hyphae and in the form and dimensions of the macroconidia. Dadant (1950), describing a disease of robusta coffee in New Caledonia, believed the pathogen to be a *Thielaviopsis*, although he saw only the phialidic state of the fungus. He suggested the name *T. neo-caledoniae* Dadant (*nom. nud.*). Orpurt and Curtis (1957) listed *Thielaviopsis sulphurellum* among the fungi isolated from prairie soils in Wisconsin. They gave no authority for the name and it does not appear to be a validly published taxon.

Berkeley and Broome (1850) described *Torula basicola* briefly as follows: "Black, effused. Hyphasma creeping, branches here and there arising from the general mass and giving off fascicles of short fastigiate fertile threads consisting of from 5-7 articulations. Articulations not constricted, ultimately separating, the last very obtuse. . . ." Their illustration of the fungus shows the conidia as segments of clavate branches separating from each other by disarticulation. No phialides or cylindrical endogenous phialoconidia were mentioned. The cotype collection IMI 742 (ex *Torula basicola* folder in K), labelled '*Thielaviopsis basicola* (Berk & Br.) Ferr., Herb. Berk. 1879 ad basim stipitum pisi, June 20, 1846,' shows only a few amber or dark brown conidia, mostly solitary but occasionally a rectilinear series of 5-6 (Figure 40A). The 'terminal segment in a series is conoid with a single transverse germ slit at its base, and the rest are short-cylindrical with one transverse germ slit at each end. No vegetative mycelium, phialides or cylindrical phialoconidia have been observed. Berkeley and Broome's characterization of the colony as 'black, effused,' suggests that their collection contained only the brown thick-walled conidia noted above. Ferraris (1910), transferring the fungus to *Thielaviopsis*, gave a revised description referring to the characters of the 'macroconidia,' the conidiophores bearing them, and the 'microconidia' originating in the interior of the hyphae. He did not mention the type specimen of *Torula basicola* nor the specimen on which he based his revised description.

A nomenclatural problem is apparent here. The name *Thielaviopsis basicola*, which is an obligate synonym of *Torula basicola*, is applicable only to the dark coloured, thick-walled spores which are the only elements present on the type specimen of *Torula basicola*. Thus both the phialidic state of the fungus, and the fungus in its entirety, are left without a name.

Examination of several later collections assigned to *T. basicola* has shown that the morphology of the phialidic state of the fungus and the formation of two different kinds of conidia closely parallels the situation in *T. paradoxa*. The only noticeable differences are:

1) The vegetative hyphae are subhyaline to pale brown, septate, branched and sometimes agregated in irregular patches;
2) The phialophores are subhyaline to pale brown, and new phialides or phialophores arise as alternate branches from below the septa in the basal part of the conidiophore;
3) The second kind of conidia are thallic-arthric, corresponding in morphology to the thick-walled spores observed in the cotype specimen of *Torula basicola*. They are in reality endo-arthroconidia, occurring as linear series of segments of clavate hyphal branches. Each of these conidia is unicellular, dark brown or amber in colour, with a thin, smooth outer wall, and a secondary inner wall 1-2 μ thick. Occasionally one or two such conidia are formed in an intercalary position (Figure 40C), the terminal and basal cells remaining hyaline and thin-walled. In cultures, while rectilinear series of 5-6 conidia are usual, occasional variations are found. In one, a thick-walled conidium may be terminal and solitary (Figure 40C) borne on a narrow, sparsely septate, hyaline conidiophore. In another, a new hyphal growth may develop from the sides of a previously terminal and solitary thick-walled conidium and, in its turn, produce a solitary, terminal, thick-walled conidium. In at least one dried agar culture examined (IMI 19856), individual cells of a few hyphal elements had also become thick-walled, dark brown, intercalary chlamydospores (Figure 40B).

Brierley (1925) described the formation of phialoconidia in *T. basicola* as resulting from longitudinal splitting of the conidiophore wall. Subsequent studies have shown that the process is, in fact, identical with the enteroblastic conidial development in *Chalara* species and *T. paradoxa*. According to Brierley (l.c.), the thick-walled arthroconidia are formed successively in the development of hyphae of limited growth. Tsao and Bricker (1970) have described the development of these arthroconidia in acropetal succession. Our own observations agree for the most part with their findings. At maturity, the thick-walled, coloured segments tend to break apart at the septa, and each is capable of germinating. Thus the development is thallic, leading to the formation of endo-arthroconidia that are capable of functioning as chlamydospores. Tsao and Bricker (1970) thought the term 'acropetal arthro-aleuriospores' appropriate for such conidia, but we prefer the terminology developed at the 1969 Kananaskis Conference on Fungi Imperfecti (see Kendrick 1971).

We believe that *T. paradoxa* and *T. basicola* are best redisposed in *Chalara*. *Thielaviopsis* thus becomes a synonym of *Chalara* (see pages 54, 60).

Chalaropsis (Figures 1, 5, 37, 38, 43B, 44, 45)

Peyronel (1916) erected the genus *Chalaropsis* Peyr., with a description which reads: "Hyphae ramosae, septatae, hyalinae vel fuscidulae; conidia dimorpha; majora subglobosa vel subelliptica, fusca, unicellularia, sessilia vel in brevibus mycelii ramis acrogena; minora cylindrica, hyalina, catenulata, ex interiore conidiophororum lageniformium generata atque ex apice eorum exsilientibus. Est *Thielaviopsis* macroconidiis simplicibus, non catenulatis," with *C. thielavioides* Peyr. as the type species. According to Mason (1941), a specimen of *C. thielavioides* was first observed as early as 1878 when Berkeley and Broome referred to a fungus on walnut shells from Scotland as *Cylindrosporium longipes* (Pr.) Cooke (in error as *Chalara longipes* Strauss). Hammond (1935), studying a *Chalaropsis* sp. on walnut, stated that she had examined Berkeley's specimen, had found on it both 'A' and 'B' conidia, and accepted it as belonging to *C. thielavioides*. McAlpine (1902) described 'A' and 'B' forms of *C. thielavioides* as two distinct fungi: *Cylindrium intermixtum* McAlpine and *Coniosporium radicicola* McAlpine. According to the International Code of Botanical Nomenclature, *Cylindrium intermixtum* provides the first valid specific epithet for this taxon. Bliss (1941) described *Ceratocystis radicicola* Bliss with a *Chalaropsis* state to which he did not give a specific epithet. Hennebert (1967) added another species, *Chalaropsis punctulata* Henneb., isolated from roots of *Lawsonia inermis* L. He noted the morphological similarities between this fungus and the *Chalaropsis* state of *Ceratocystis radicicola,* but considered *C. punctulata* distinct because of differences in the dimensions of 'aleuriospores' and phialoconidia. Sugiyama (1968) described a new variety of *C. thielavioides* under the varietal name *ramosissima* Sugiyama and considered it distinct from var. *thielavioides* in its faster growing colony, larger 'aleuriospores,' and particularly in its branched phialophores. He also stated that the new variety differed from the *Chalaropsis* states of *Ceratocystis radicicola* and *Ceratocystis variospora* (Davids.) C. Moreau mainly in the length and branching of the phialophores and the size of the conidia.

The type specimen of *C. thielavioides* in Saccardo's herbarium in PAD bears only conidia (Figure 44A), but nevertheless conidia of two types: a) cylindrical conidia with blunt or rounded ends, unicellular and hyaline; b) subglobose, ellipsoidal or obovoid, unicellular brown conidia lacking vertical or transverse germ slits.

In a number of specimens assigned to this species, the following features are apparent:

The vegetative hyphae are septate, branched, and hyaline to subhyaline.

The conidiophores are of two types:

1) Simple, hyaline, subhyaline or pale brown phialophores, arising as erect branches of the vegetative hyphae. Their walls are smooth or minutely verrucose. The phialide, terminating the conidiophore, is subcylindrical to obclavate, composed of a subcylindrical or ellipsoidal venter and a narrow, cylindrical collarette, concolorous with the rest of the hyphae. Transition from venter to collarette is

gradual. New phialides or phialophores arise as lateral outgrowths from the basal cells of existing phialophores.

2) Hyaline, short, aseptate or scantily septate hyphae bearing solitary, terminal, holoblastic conidia. Such conidiophores often branch sympodially.

The conidia borne in the phialides are enteroblastic, cylindrical with almost rounded or truncate ends, unicellular, hyaline, subhyaline or pale brown and extruded in long basipetal chains from the open apex of the collarette. The conidia of the second type are holoblastic, globose, subglobose, ellipsoidal, ovoid or pyriform with truncate or blunt base, have smooth or minutely verrucose walls, and are unicellular and yellowish brown to brown. No germ pores or slits have been seen on these conidia.

A viable culture of *Chalaropsis thielavioides* was not available for our study of conidium ontogeny. However, herbarium material of the fungus bearing conidiophores in different stages of growth shows that the phialoconidia in this fungus develop in the same manner as those of *Chalara* spp. and *T. paradoxa*. The holoblastic, dark brown, thick-walled conidia appear to arise just as do similar conidia in *T. paradoxa* (Figure 8, type III). Observations carried out on the ontogeny of conidia in *Ceratocystis radicicola*, which has a *Chalaropsis* state, have shown that it produces enteroblastic-phialidic conidia like those of *T. paradoxa*. Some of the cylindrical phialoconidia also become dark and acquire thick walls (Figure 5H, I) as in *T. paradoxa*. Development of the thick-walled, darkly pigmented conidia appears to be predominantly holoblastic (Figure 5A-F). Figure 5A, at the start of the sequence, shows a hyphal branch that has produced a solitary, terminal, thick-walled, dark conidium, and two new branches beneath it. The lowermost shows a slight bulbous swelling at its apex. In Figure 5B, 60 minutes later, the swollen apex is recognizable as a conidium initial which attains full size over a period of 180 minutes (Figure 5E). The conidium is separated from the conidiophore by a septum and begins to darken and acquire thick walls (Figure 5C-E). Figure 5F shows the final appearance of the conidium.

We treat *Chalaropsis* as a synonym of *Chalara*. See pages 54, 60.

Hughesiella (Figure 42C)

Batista and Vital (1956) proposed *Hughesiella* Bat. & Vital, for a dematiaceous fungus isolated from air. The genus was characterized as having hyaline conidiophores of two forms—short phialides with long 'catinules' of ellipsoid, continuous, brownish-black conidia; and simple to branched elongate hyphae producing unicellular, lenticular, blackish conidia with an equatorial, subhyaline band, single or catenulate. The authors noted the affinity of the genus with *Chalaropsis*, but considered it distinct from the latter because of its brown phialoconidia. The genus has remained monotypic with *H. euricoi* Bat. & Vital as the type species.

DAOM has the original type culture ex IMUR 640 in the form of dried agar discs derived from 13-day-old cultures grown on Sabouraud's agar. Vegetative hyphae are hyaline, septate, sparsely branched and smooth-walled. Conidiophores arising as lateral branches of the hyphae are frequently indistinguishable from the latter. They are simple, hyaline to subhyaline, cylindrical, and terminate in a phialide which is subcylindrical to obclavate. The transition from venter to collarette is gradual.

The conidia are of two kinds. The first kind are enteroblastic-phialidic, cylindrical, ovoid or doliiform, with blunt or somewhat rounded ends, unicellular (Figure 42C), hyaline to subhyaline, and extruded from the open end of the collarette in long, easily dispersible chains. The second kind of conidia are short-cylindrical, oblong, ovoid or occasionally pyriform with a truncate base, unicellular, yellowish

5) *Chalara* state of *Ceratocystis radicicola*. A-F. 35 mm time-lapse sequence of chlamydospore formation at 0, 60, 125, 180 and 770 minutes respectively. G. Sympodial development, each branch terminating in a single chlamydospore; H-I. Changes occurring in chains of phialoconidia.

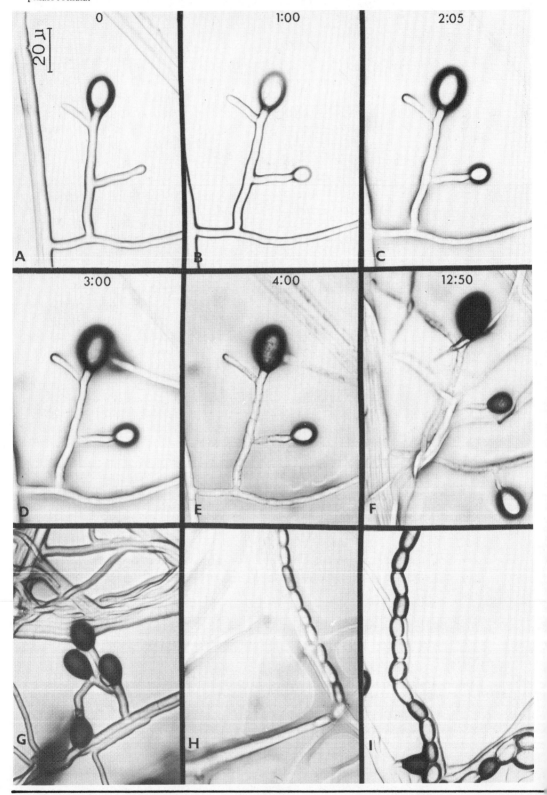

brown or reddish brown, with an oblique or vertical germ slit, and are frequently borne on simple, short, narrow, sympodially branching conidiophores. Dark brown thallic conidia, continuous or sparsely septate, sometimes resembling young phialides or hyphal cells, are also seen in the cultures in fairly large numbers.

In a 35-day-old dried PDA culture of this isolate, the phialophores are 0-1-septate, hyaline, and terminate in a subcylindrical phialide composed of a subcylindrical or ellipsoidal venter and a narrow cylindrical collarette. The conidia are virtually identical with those formed on Sabouraud's agar. Some of the thick-walled conidia have germinated, and the germ tubes, functioning as secondary phialophores, have become thick-walled and dark brown, producing cylindrical, hyaline, thin-walled conidia in short chains at their apex.

Dr. Sequeira of URM kindly sent us a living culture ex IMUR 640. The morphology of its vegetative hyphae, phialides and the conidia conform to the description given above, and its conidium ontogeny parallels that seen in *Thielaviopsis paradoxa*. We consider *Hughesiella* a synonym of *Chalara*. See pages 55, 60.

Stilbochalara (Figure 42B)

Ferdinandsen and Winge (1910) published *Stilbochalara* Ferd. & Winge with the type species *S. dimorpha* Ferd. & Winge, collected on semi-putrid fruits of *Theobroma cacao* L. in Las Trincheras, Venezuela. The generic diagnosis was brief: "Genus phaeostilbeum, conidia endogena Chalarae modo gingens—est *Chalara* stilbiformis." The specific epithet was apparently derived from the two kinds of conidia that they observed in this fungus. They wrote: "as appearing from the diagnose the brown conidia can be found in the same tubuli as the hyaline ones (and often presenting an intermediate aspect) quite as in *Thielaviopsis*; especially, however, their formation is confined to tubuli near the basis of the coremium, i.e., as far localized. This fact suggests—coupled with the thickening of the walls in the brown conidia—that the dimorphism in this species is normal or at least being about to become established." Ainsworth and Bisby (1943) stated that the type specimen had been seen and that the name was synonymous with *Thielaviopsis*, although they did not indicate the appropriate combination or synonymy.

The holotype specimen comprises a few pieces of dried plant material and is in poor condition, bearing only sparse fungal elements. A few *Ceratocystis* perithecia possessing long, brownish-black necks with hyaline to subhyaline, non-septate ostiolar hyphae that are broad at the base and gradually tapering towards the apex, have been observed. Numerous irregularly shaped, lobed appendages corresponding to those described by Dade (1928) as typical of *Ceratocystis paradoxa* are present on the walls of the perithecial venters. Ascospores have not been seen.

The few unbroken phialophores that can be found are simple, subhyaline to brown, many-septate with smooth walls. The phialide terminating the phialophore is subcylindrical to conical, with a cylindrical or subcylindrical venter and a narrow cylindrical collarette. The transition from venter to collarette is gradual. Occasionally, several conidiophores form loose clusters or lax coremia.

The conidia are of two kinds. The first kind are enteroblastic-phialidic, cylindrical, ovoid or doliiform with blunt or truncate ends (Figure 42B), unicellular, hyaline, subhyaline or pale brown, with smooth walls. The second kind are ovoid with blunt or obtuse ends, unicellular, reddish brown to dark brown, with smooth walls and vertical germ slit, frequently borne on simple, short, hyaline conidiophores. Conidium ontogeny is identical to that seen in *Thielaviopsis paradoxa*.

We consider *Stilbochalara* a synonym of *Chalara*. See pages 55, 60.

Conidial states of some Ceratocystis species and Cryptendoxyla hypophloia

Ceratocystis
Several species of *Ceratocystis* Ellis & Halst. are known to have conidial states in *Chalara, Thielaviopsis* and *Chalaropsis* as follows:

1) Species with reported conidial states in *Chalara*:
   A) *Ceratocystis coerulescens* (Münch) Bakshi: conidial state–*Chalara ungeri* Sacc.
   B) *Ceratocystis fagacearum* (Bretz) Hunt: conidial state–*Chalara quercina* Henry.
2) Species with reported conidial state in *Thielaviopsis*:
   Ceratocystis paradoxa (Dade) C. Moreau: conidial state–*Thielaviopsis paradoxa* (de Seynes) Höhn.
3) Species with reported conidial state in *Chalaropsis*:
   Ceratocystis radicicola (Bliss) C. Moreau: conidial state–without a specific epithet.
4) Species recognized as having an 'endoconidial' imperfect state*
   Ceratocystis adiposa (Butl.) C. Moreau;
   Ceratocystis antennaroidospora Roldan *nom. illegit.;*
   Ceratocystis asteroides Roldan *nom. illegit.;*
   Ceratocystis autographa Bakshi;
   Ceratocystis filiformis Roldan;
   Ceratocystis fimbriata Ell. & Halst;
   Ceratocystis major (Van Beyma) C. Moreau;
   Ceratocystis moniliformis (Hedgc.) C. Moreau;
   Ceratocystis variospora (Davids.) C. Moreau;
   Ceratocystis virescens (Davids.) C. Moreau.

The morphological features of *Chalara quercina* and *C. ungeri* have been considered earlier. *Ceratocystis major* is known to have a *Graphium* state and *C. variospora* has been considered a synonym of *C. fimbriata* by Webster and Butler (1967). The following account refers to the conidial states of the other *Ceratocystis* species listed above.

Ceratocystis paradoxa (Figure 42A). A slide prepared from the type specimen has been examined. The phialophores are generally indistinguishable from the vegetative hyphae till the liberation of conidia begins. The general morphology and dimensions of the phialides and of the dimorphic conidia match those of *Thielaviopsis paradoxa*. A good *Chalara*.

Ceratocystis radicicola (Figure 38). The vegetative hyphae are septate, hyaline to pale brown, and sparsely branched. The phialophores are simple, cylindrical, septate, hyaline to subhyaline or pale brown. The phialides are subcylindrical to lageniform, composed of a subcylindrical venter and a cylindrical collarette; transition from venter to collarette is gradual or often imperceptible. The conidia are of two kinds. The first are enteroblastic-phialidic, cylindrical with blunt or rounded ends, unicellular, hyaline to subhyaline, occurring singly or in easily dispersible chains. The second kind are ovoid or pyriform with a truncate base, unicellular, brown to dark brown, solitary and acrogenous on short, septate, hyaline conidio-

---

* To the best of our knowledge, no other *Ceratocystis* species described since 1962 has a conidial state in *Chalara, Thielaviopsis* or *Chalaropsis*.

phores that proliferate sympodially (Figure 5G). Some of the cylindrical phialoconidia become dark and thick-walled (Figure 5H, I), a feature reminiscent of *T. paradoxa*. Proliferation of conidiophores is sympodial (Figure 5G) reminiscent of that
occurring in *T. paradoxa* and *T. basicola*. A good *Chalara*.

Ceratocystis adiposa (Figure 37). The vegetative hyphae are hyaline at first, darkening with age, septate, irregularly branched, with thin, smooth or somewhat verrucose walls constricted at the septa. The phialides are subcylindrical to lageniform,
composed of a subcylindrical or ellipsoidal venter and a cylindrical collarette; transition from venter to collarette is gradual or sometimes abrupt. Two kinds of
conidia are present: 1) enteroblastic-phialidic, cylindrical to doliiform, with truncate or almost rounded ends, or ovoid or pyriform with a truncate base, unicellular,
hyaline, subhyaline or pale brown with smooth or verrucose walls; 2) globose,
ovoid or oblong, unicellular, brown to reddish brown, with thick walls which are
verrucose, or covered with papillae, or fimbriate; germ slits obscure. Conidia of the
second kind occur in short chains or singly on short, narrow, cylindrical, septate,
subhyaline to pale brown conidiophores. A good *Chalara*.

Ceratocystis antennaroidospora Roldan. *nom. illegit.* and *Ceratocystis asteroides*
Roldan. *nom. illegit.* The status of these two species is discussed on page 171.

Ceratocystis filiformis Roldan. We have considered this as a synonym of *Ceratocystis moniliformis*.

Ceratocystis fimbriata (Figure 45). The vegetative hyphae are subhyaline to pale
brown, septate, branched, and smooth-walled. The phialophores arise as lateral
branches of the vegetative hyphae and are simple, cylindrical, 1-5-septate, subhyaline to pale brown at the base, becoming progressively lighter toward the distal end,
with smooth walls occasionally constricted at the septa. The phialides are subcylindrical to lageniform, composed of a cylindrical, subcylindrical or ellipsoidal venter
and a cylindrical collarette; transition from venter to collarette gradual. Two types
of conidia are present: 1) enteroblastic-phialidic, cylindrical with blunt or truncate
ends, unicellular, hyaline, subhyaline or pale brown, and smooth-walled; 2) globose, subglobose, oval, doliiform or pyriform with truncate base, unicellular, brown
to dark brown, with smooth, thick walls. Germ slits are not apparent. The dark,
thick-walled conidia occur in short chains or singly on cylindrical, septate or aseptate, subhyaline to pale brown conidiophores. A good *Chalara*.

Ceratocystis moniliformis (Figure 43A). Vegetative hyphae are hyaline, septate and
branched. Phialophores cylindrical, 1-2-septate, or often reduced to conidiogenous
cells, hyaline with verrucose or sometimes smooth walls. Phialides subcylindrical to
lageniform with a gradual transition from venter to collarette. The conidia are
enteroblastic-phialidic, cylindrical with blunt or truncate ends, unicellular, hyaline,
occurring in short chains. Long chains of doliiform, hyaline or pale brown phialoconidia have been described by other workers. A good *Chalara*.

Roldan (1962) described and illustrated *Ceratocystis filiformis* Roldan from cultures isolated from three hosts: *Calamus maximus, Endospermum peltatum* and
Parkia javanica. He neither designated a type for the name, nor apparently deposited a type culture in a culture collection. Roldan indicated the affinity of his
isolate with *C. moniliformis* but considered his fungus distinct because it had longer
ornamental perithecial hyphae and endoconidia than *C. moniliformis*. The length of

the conidia as well as the length of the ostiolar hyphae reported by Roldan appear to fall well within the broad limits of *C. moniliformis*, with which it is placed in synonymy.

Ceratocystis autographa (Figure 28C). Bakshi's (1951) description and illustration of the fungus leaves no doubt that it is a good *Chalara*, and a study of slides made from the type by Mr. H. P. Upadhyay has confirmed this. The phialoconidia are short clavate rather than barrel shaped. The round or ovoid conidia presumably develop in a manner comparable to type II of *T. paradoxa*, a feature that needs to be confirmed by a study of a viable culture of the fungus, to which we had no access.

Ceratocystis virescens. For a discussion of this species see page 138.

Cryptendoxyla (Figure 30A)
Cryptendoxyla hypophloia. Malloch and Cain (1970) described and illustrated *Cryptendoxyla hypophloia,* the only member of the genus *Cryptendoxyla* Malloch and Cain, with an asexual state which is clearly a *Chalara*. Through the courtesy of Dr. Malloch, we have been able to examine authentic cultures of the fungus. Vegetative hyphae are hyaline to light brown, septate and branched. Phialophores are 1-septate at the base, hyaline and smooth-walled. Phialides are subcylindrical with gradual or barely perceptible transition from venter to the long collarette. The conidia are enteroblastic-phialidic, cylindrical with blunt ends, unicellular, hyaline, and occur in chains. A good *Chalara*.

Excioconidium (Figure 15)
Plunkett (apud Stevens 1925) erected the monotypic genus *Excioconidium* Plunk., with *E. cibotti* as the type species, collected on *Cibotium chamissoi* in Hawaii. A short generic description was given: "Fertile hyphae erect, dark, septate, conidia hyaline, septate, cylindric, borne internally in fertile hyphae," and an illustration provided. The genus was considered to be distinct from *Chalara* on the basis of spore septation. Clements and Shear (1931) listed it as *Excioconis* Plunk.

In the isotype specimen and another collection of the fungus, the vegetative mycelium is immersed in the substrate. Aggregations of short, broad, dark brown, thick-walled hyphal cells arranged in thin, flat, subcuticular layers from which phialophores originate, are often seen. The phialophores are simple, usually gregarious, but sometimes solitary and scattered, erect, cylindrical, 3-4-septate and terminate in a phialide. The wall is smooth, thick and dark reddish brown in the basal part, but becomes progressively thinner and lighter toward the apex of the phialide. The phialides are of the urceolate, obclavate or subcylindrical type described for *Chalara* and have a cylindrical to ellipsoidal venter and a cylindrical to obconical collarette; the transition from venter to collarette is gradual.

The conidia are enteroblastic-phialidic (phialoconidia), cylindrical to ellipsoidal with the apex usually rounded and the base truncate with a distinct marginal frill (Figure 15); hyaline, mostly 7-septate (occasionally 5-septate). The conidium walls are smooth and without constrictions at the septa. Occasionally more than one conidium is seen within the collarette.

Both morphology and conidium ontogeny exactly follow the pattern set by *Chalara* spp. and we treat *Excioconidium* as a synonym of *Chalara*. See pages 55, 60.

Fusichalara (Figure 46)
The generic name *Fusichalara* was proposed by Hughes and Nag Raj (1973) to accommodate ten collections of *Chalara*-like fungi found in New Zealand. The authors noted the marked similarity of *Fusichalara* to *Chalara* but considered it

distinct by virtue of the thickened inner wall of the phialide at the zone of transition from venter to collarette, and the formation of two kinds of multiseptate, hyaline or coloured conidia from the same phialide. The first-formed conidia are cylindrical, straight, and about twice as long and with twice as many septa as the subsequent conidia, which are fusiform and straight or sigmoid. Three species were recognized.

We accept *Fusichalara* as a good genus. See pages 55, 144.

Sporendocladia (Figure 54)

Arnaud (1954) published the generic name *Sporendocladia* Arn. (*nom. nud.*) with *S. castaneae* Arn. (*nom. nud.*) as the single species, for a fungus occurring on cupules of *Castanea*. He gave a brief generic description "Ampoules sporifères groupées en touffe au sommet d'un conidiophore brun, simple ou peu ramifié; le reste comme *Chalara* et *Cylindrotrichum*." The brief description of *S. castaneae* was as follows: "vieilles cupules de *Castanea*; (G.A. no. 2230). . . . Conidiophores bruns, de 70-150 μ de haut, avec au sommet 6 á 8 ampoules sporifères - articulés - spores incolores, de 3 x 1 μ." Arnaud's beautiful illustration shows a single, darkly pigmented conidiophore bearing a terminal cluster of lageniform phialides, each of which produces a chain of enteroblastic-phialidic conidia that are cylindrical with truncate ends, unicellular and hyaline. We have examined the original specimen of *S. castaneae* (no. 2230 in PC) and found (as we expected) that Arnaud's observations were accurate.

We accept *Sporendocladia* as a genus distinct from *Chalara* and have validated *Sporendocladia** and *S. castaneae** on page 162.

Sporoschisma (Figures 52, 53)

A detailed account of *Sporoschisma* Berk. & Br. (established in 1847 by Berkeley) was published by Hughes (1949). In a subsequent paper, describing species of *Sporoschisma* found in New Zealand, Hughes (1966) assigned the known perfect states of two species to *Chaetosphaeria* Tul. [These are now disposed under *Melanochaeta*.]

As delimited by Hughes (1966), the genus *Sporoschisma* is characterized by massive, dark brown, stalked, cylindrical phialides [phialoconidia 3-5-septate, cylindrical, pale brown to dark brown, end cells sometimes paler than the rest] intermixed with much smaller and narrower, erect, sterile, apically swollen capitate hyphae that bear a cap of (?) mucilage. Hughes recognized four species.

We have nothing significant to add to the observations and descriptions published by Hughes (1949, 1966), which, in fact, form the basis for our section on *Sporoschisma* (see page 157).

Chaetochalara (Figures 47-51)

Sutton and Pirozynski (1965) proposed the generic name *Chaetochalara* Sutton & Piroz. to accommodate three species: *C. bulbosa* Sutton & Piroz., *C. africana* Sutton & Piroz. and *C. cladii* Sutton & Piroz., with *C. bulbosa* as the type species. These authors recognized the affinity of their genus with *Sporoschisma* and *Chalara*, and suggested that it was intermediate between the two in taxonomic position. They thought that the sterile capitate hyphae occurring in *Sporoschisma* could be considered analogous to setae, but that the large size of the phialides and the multiseptate conidia of *Sporoschisma* clearly separate it from *Chaetochalara*. They considered *Chaetochalara* close to *Chalara* in phialide and conidium development, but distinct from it, since neither setae nor any form of associated sterile elements occur in the

---

*   Very recently Sutton (in Trans. Br. mycol, Soc. 64: 411, 1975) has treated *Sporendocladia castaneae* as a synonym of *Phialocephala fumosa* (Ell. & Ev.) Sutton.

latter. In advancing this opinion, it would appear that they overlooked Harkness's *Chalara setosa*. Pirozynski and Hodges (1973) added a fourth species, *C. aspera* Piroz. & Hodges, to the genus.

Type specimens of the four species of *Chaetochalara* and other unnamed collections have been examined. According to Sutton and Pirozynski (1965) the immersed vegetative mycelium is composed of hyaline hyphae often aggregated in substomatal cavities and emerging through stomates to form the superficial mycelium, as seen on the specimens of all species but one. This mycelium is usually composed of subhyaline to pale brown, septate and branched hyphae, from which the setae and phialides arise. In one species, the superficial hyphae often form thin, prosenchymatous layers of thick-walled, subhyaline, short, broad, cylindrical cells from which simple or irregularly branched, short phialophores arise (Figure 51), each branch terminating in a phialide.

The phialophores are often represented by phialides borne directly on the superficial hyphae, but in one species each conidiophore is simple, erect, cylindrical, and up to 4-septate with a slightly swollen basal cell. The walls are almost always smooth, thick, and darker in the basal supporting structure. However, in *C. aspera* they are asperate. The phialides closely resemble those described for *Chalara*, though the urceolate configuration has not been seen in *Chaetochalara*. As in *Chalara*, the transition from venter to collarette can be abrupt, gradual, or barely perceptible. The wall is of even thickness in most species, but in *C. aspera* it gradually becomes thinner toward the apex. Wall ornamentation is found only in *C. aspera* where the collarette is sparsely echinulate. The phialide wall is subhyaline, pale brown, yellowish brown or fuliginous and usually concolorous except near the collarette.

Infrequent percurrent or sympodial proliferation of the conidiophore occurs in one species, and the percurrently proliferated conidiophores can be mistaken for sterile setae.

The conidia are enteroblastic-phialidic (phialoconidia), cylindrical, with rounded or blunt ends or, in some species, the apex is rounded and the base truncate with or without a marginal frill. They are unicellular, 1-septate or 1-2-septate, the cells often of unequal length, and invariably hyaline. The walls are smooth, not constricted at any septa present, usually thin but relatively thicker in two species. Conidium size varies with the species, but large, thick-walled, multi-septate conidia like those of *Chalara insignis* or *Chalara dictyoseptata* have not been found in this group. The conidia are extruded from the open end of the collarette singly or in short chains. Thick-walled, coloured conidia were not observed in any species of *Chaetochalara*.

Sterile setae occur interspersed with the phialophores. They are generally not very numerous (sparse in one species), and are simple, erect or variously bent, but extending to some distance above the phialides or phialophores. They generally resemble the phialides, but have darker colour, thicker walls, septa, a blunt or pointed apex, and are, of course, sterile. In *Chaetochalara africana*, the setae are nodulose.

We maintain *Chaetochalara* as a genus distinct from *Chalara*. See pages 55, 148.

Ascoconidium (Figures 55, 56)

Seaver (1941) established the genus *Ascoconidium* Seaver with *A. castaneae* Seaver as the only species, occurring on *Castanea dentata*. Because of the morphological similarity of the conidiophores to asci, he misguidedly, in our opinion, introduced the terms 'ascoconidiophores' and 'ascoconidia.' He stated: "This fungus may be transferred to the genus *Endoconidium* Prill. & Delacr. but differs from it in that there is only a single 3-septate spore in each ascoconidiophore and in that the ascoconidiophore is more ascus-like." Funk (1966a) redescribed the genus after a study of the original specimen and considered that *Ascoconidium* was morphologically close to *Chalara* and *Excioconidium*. Later (1966b) he added another species, *A. tsugae* Funk, to the genus. Funk (1975) has established that *A. tsugae* is the conidial state of a Helotiacious discomycete, *Sageria tsugae* Funk.

We have examined the type specimens of both species. Both possess a pseudoparenchymatous basal layer, subepidermal in *A. tsugae*, but innate and subsequently erumpent in *A. castaneae*.

The phialophores arise from the terminal layer of cells of the basal stroma and may comprise a narrow, short, cylindrical supporting cell and a phialide; or a supporting structure of one to several cells may subtend a phialide. The phialides are clavate to subcylindrical with a blunt or rounded apex, brown to dark brown, and smooth or rough walled. The walls of the phialide and the supporting structure are of uniform thickness.

The conidia are enteroblastic-phialidic, cylindrical, with a slightly rounded or blunt apex and a truncate base with a conspicuous marginal frill; hyaline, with smooth, thick walls. The conidia in *A. castaneae* are 3-septate, while in *A. tsugae* they are muriformly septate with 7 transverse and 1-4 oblique or longitudinal septa. Funk (1966b) grew *A. tsugae* in culture, but could not observe sporulation.

Conidium ontogeny in *Ascoconidium* as noted by Funk (1966a), is similar to that of *Sporoschisma* and *Chalara* except for one feature. A mature, well-differentiated conidium occupies the major part of the phialide. The basal wall of the conidium is in contact with a bulging cytoplasmic protuberance that is the next conidium initial. With the development of the second conidium the first mature conidium is pushed against the apical, thick, inelastic wall of the phialide. Increasing pressure from below presumably causes a longitudinal split in the apical part of the phialide and the mature conidium is extruded. After the release of the conidium, the lacerated flaps of the phialide wall appear to fall back into place, seemingly closing the phialide again (Figure 56). In *Chalara*, the phialide opens by a circumscissile split in its apical wall.

We recognize *Ascoconidium* as a distinct genus. See pages 56, 164.

Bloxamia and Endosporostilbe (Figures 6, 7, 57)

Berkeley and Broome (1854) published the generic name *Bloxamia* Berk. & Br., based on *B. truncata* Berk. & Br. (Figure 6), a tuberculariaceous fungus occurring on dead, decorticated wood in England. The generic description was: "Peridium deorsum persistens, sursum delicatissimum hyalinum evanescens demum excipuliforme; sporidia quadrate tubulis arcte congestis enata. . . ." Pirozynski and Morgan-Jones (1968) redescribed *B. truncata* from the type specimen and listed its synonyms. Höhnel (1910) transferred *Catinula leucophthalma* Lev. to *Bloxamia*, forming the combination *B. leucophthalma* (Lev.) Höhn.

A study of the type and other collections of *B. truncata* has shown that the fructifications are scattered or gregarious, black, disciform sporodochia with pale brown, superficial stromata composed of long, subhyaline to pale brown, smooth-walled cells arranged in dense palisades. The conidiophores arise from these cells, and each consists of a long, septate, supporting structure subtending a phialide. The phialide is cylindrical to subcylindrical with a morphologically undifferentiated collarette and smooth walls. The conidia are enteroblastic-phialidic, cuboid to short-cylindrical, rarely oblong with truncate ends, unicellular, and hyaline to subhyaline. Pirozynski and Morgan-Jones (1968) noted and illustrated percurrent proliferation of phialophores in this fungus. Our examination of *B. leucophthalma* and *B. tetraploa* Berk. & Br., has shown that these are conspecific with *B. truncata*.

Subramanian (1958) described *Endosporostilbe* Subram., type species *E. nilagirica*, on dead twigs amongst litter in Ootacamund, India, with one-celled, endogenous conidia formed in basipetal succession from the apex of conidiophores aggregated into synnemata. The genus has remained monotypic. Illman (1964) considered the name superfluous, and suggested that the fungus was congeneric with *Stilbochalara dimorpha*.

6) *Bloxamia truncata* [Redrawn from Berkeley and Broome 1854]

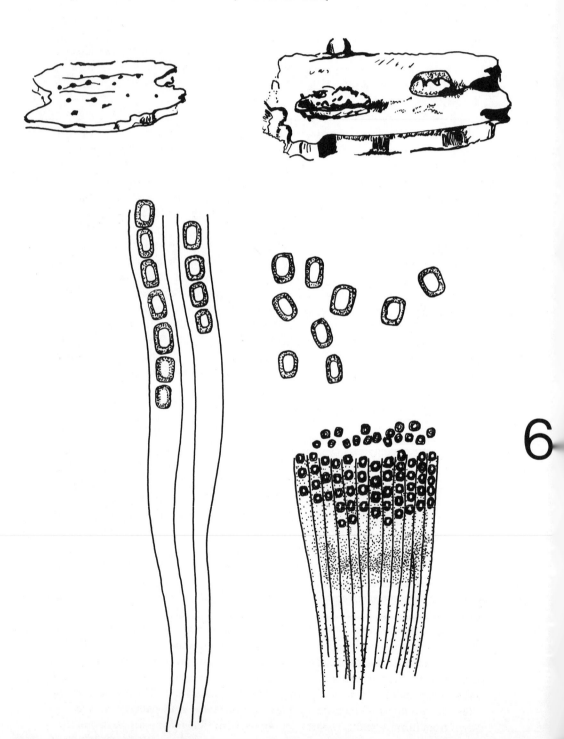

6

We have examined a slide prepared from the type specimen showing a single, more or less cylindrical synnema terminating in a conidiogenous head. The synnema (Figure 7) is composed of subhyaline to pale brown, simple, septate, closely aggregated hyphae. The hyphae of the synnema widen gradually and become darker, forming the core of the synnematal head. The terminal cells of these hyphae function as phialides, and are cylindrical to subcylindrical with thin, smooth walls. The conidia, enteroblastic-phialidic as in *Chalara, Bloxamia* etc., are short- or long-cylindrical with truncate ends, unicellular, and hyaline, with thin, smooth walls. Successive basipetally produced conidia remain together in chains, and the profuse conidia formed from the numerous phialides accumulate in a slimy mass over the head. The fungus has not been grown in pure culture.

We consider *Endosporostilbe* a synonym of *Bloxamia*, and recognize the latter as a distinct genus. See pages 56, 167.

Conidium development in *Hughesiella, Stilbochalara, Ceratocystis paradoxa, Ceratocystis adiposa* and *Ceratocystis fimbriata* follows the patterns set by *T. paradoxa*. In *Ceratocystis adiposa* the short collarettes involved in Type II of *T. paradoxa* are not easily discernible. Andrus and Harter (1933) have demonstrated the 'endogenous' nature of the thick-walled, dark brown conidia of *Ceratocystis fimbriata*. We have observed that, in a very late phase of growth, this fungus produces such conidia in a holoblastic manner as well, analogous to type III of *T. paradoxa*.

The Schematic diagram in Figure 8 summarizes the types of conidium development found in *Chalara, Chaetochalara, Excioconidium, Fusichalara, Sporendocladia, Bloxamia,* and *Ceratocystis moniliformis* (I); *Thielaviopsis paradoxa, Ceratocystis paradoxa, Ceratocystis adiposa, Ceratocystis fimbriata, Hughesiella,* and *Stilbochalara* (I, II, III); *Thielaviopsis basicola* (I and IV); *Chalaropsis* and *Ceratocystis radicicola* (I and III).

Milowia Massee

Massee (1884) described *Milowia* as follows: "Pulvinate, monoecious. Mycelium sparsely septate, branched, flexuous, giving origin to numerous lateral fertile three-celled branches. Pollinodium clavate, springing from the basal cell of the fertile branch. Carpogonium formed from the terminal cell of the fertile branch, broadly obovate, producing from near its apex from 2-5 cylindrical octosporous asci." *Milowia nivea*, the single species of the genus, was described as follows: "Tufts globose, minute, white; sporidia colourless, cylindrical, truncate; conidia globose, moniliform, occupying the same position as the carpogonium when the latter is not developed." Cook (1885) remarked: "The question which suggests itself in connection with this fungus is, as to the nature of the naked clavate bodies, termed asci, viz., whether they are merely more completely developed hyphae with dissilient joints, as in *Sporoschisma*, or asci, as in Ascomyce[te]s." Massee (1893) gave an illustration and revised description of the fungus as follows: "Tufts globose, 1 line across, snow white; the erect hyphae containing the conidia 60-70 x 7-8 μ and containing 6-8 cylindrical, abruptly truncate conidia, 9-10 x 6-7 μ." He also remarked that *Milowia* is a counterpart of *Sporoschisma*, and that the conidia present the appearance of spores in an ascus. Massee's collection was thought to be in NY, but Dr. C. T. Rogerson could not trace it there, and informed us that only Massee's drawing of the fungus is available. Attempts to locate Massee's specimen in other herbaria have been futile. Massee's description and illustration are of little use in interpreting his concepts of the genus and its type species. The first description refers to two kinds of conidia: 'cylindrical, truncate, colourless sporidia' and 'globose, moniliform conidia,' but the second description refers only to 'cylindrical, abruptly truncate conidia.' Several questions arise when we try to analyse these

7) *Endosporostilbe nilagirica* [After Subramanian 1958]

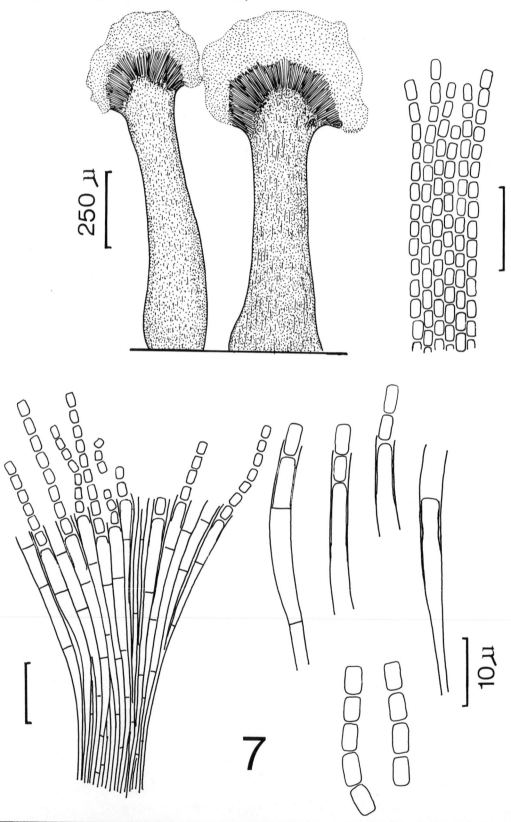

descriptions. What is the nature of the globose, moniliform conidia mentioned in the first description? Could these conidia be pigmented and thick-walled? Does the second description deliberately exclude the globose moniliform conidia altogether, or is it just a clarification of the nature of the cylindrical conidia? In other words, does the fungus have only one conidial state, in the form of erect, cylindrical conidiophores whose contents develop into conidia that escape through the ruptured apex; or does it have two different conidial states? These questions are necessary because Wakefield and Bisby (1941) assigned *Milowia nivea* to *Thielaviopsis basicola*, an important plant pathogen, though we do not know whether they had access to Massee's specimen. The true identity of this fungus can be resolved only by a study of the original specimen, and *Milowia nivea* must therefore, for the present, be regarded as a *nomen dubium*.

Endoconidium

Prillieux and Delacroix (1891) proposed the generic name *Endoconidium* Prill. & Delacr. with the following description: "Sporodochia pulvinata, albida, sporophoris hyalinis, ramosis; conidia hyalina, rotundata, in interiore ramulorum subinde generata et mox apice exsilientia." *E. temulentum* Prill. & Delacr., occurring as a parasite on rye seeds, was described as the single species of the genus with: "Mycelium hyalinum, sub superficie grani effusum, stromatice intricatum, primum extra inconspicuum, dein pulvinula initio candida, dein lenissime rosea 1/2-1 mm. 1/2 lata producens; sporophoris hyalinis septatis, guttatis, subtortuosis, 3 μ latis, bis terve repetito ramosis; conidia hyalina, e spherico ovoidea, in interiore ramulorum sporophori catenulatim nascentia, dein libera, 2.5 μ circiter." In 1892 Prillieux and Delacroix gave the name *Phialea temulenta* Prill. & Delacr. to the perfect state of the fungus, which is now regarded as *Gloeotinia temulenta* (Prill. & Delacr.) Wilson, Noble & Gray. Malcolm *et al.,* (1954) considered *E. temulentum* as the 'microconidial state' with another 'macroconidial state' also figuring in the life cycle of the fungus. Arnaud and Barthelet (1936) gave detailed descriptions of the microconidial state and the apothecia of *Sclerotinia pseudotuberosa* Arn. & Barth. They considered the 'microconidial state,' to which they gave the generic name *Rhacodiella*, with *R. castaneae* as the single species, only slightly distinct from *Endoconidium*.

Whetzel (1937) wrote: "Since the so-called microconidia of the Discomycetes as well as those of Pyrenomycetes and the lichen fungi doubtless function in fertilization, the writer proposes to revert to the original appellation for these cells, viz. 'spermatia.' This term is to be preferred to 'microconidia' and its derivatives in spite of the fact that the spermatia in some forms and under certain conditions function as true conidia." Whether form-genera are necessary to accommodate spermatial states is questionable. If the spermatial state of the fungus is to be interpreted as an asexual state, despite its potential sexual function, the Code of Botanical Nomenclature would not, regrettably, prevent the provision of a binomial for such a state. It appears, therefore, that the name *Endoconidium* may be applied to the spermatial states of a few members of the Sclerotiniaceae. It is curious that the true conidial state of this fungus should have no separate binomial.

Patouillard and Lagerheim (1891) added a second species, *E. ampelophilum* Pat., to the genus, with the following circumscription: "Taches circulaires. . . . Tubercules convexes, dimidiés, 4-100 μ de diam., d'abord sous epidermiques puis libres, composés de files de conidies incolores où un peu jaunâtres. Chaque file comprend un filament cylindrique, attenué à la base en un stripe tres-court et largement ouvert à son sommet, renfermant 4-7 conidies libres; celles-ci sont rondes où ovoïdes et mesurent 4-5 μ. Les filaments conidifères sont simplices et se séparent difficilement les uns des autres. Sur des raisins murs. Ambato." They did not designate a holotype. The Patouillard collection in FH has a specimen labelled

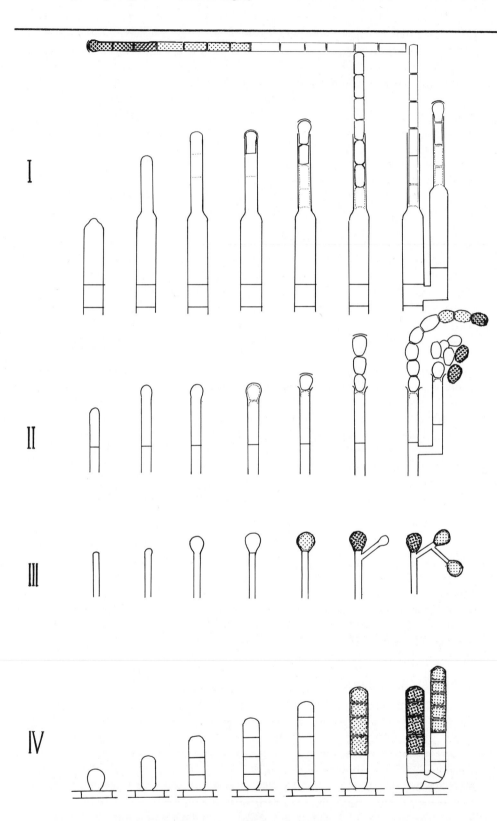

"*Endoconidium ampelophilum* sur des raisins rendoi, Ambato, Mars 1891, comm. de Lagerheim, Patouillard?" This is apparently the original specimen. The packet contains three dried fruit skins (?) on none of which was a fungus colony visible. Scrape mounts showed only a few hyphal fragments. No spermatia, phialides or phialoconidia were seen. Since the material bears no elements that could reasonably be referred to *Endoconidium, E. ampelophilum* is here considered a *nomen dubium*.

Delacroix (1893) described and illustrated two more species: *E. luteolum* and *E. fragrans.* He considered *E. fragrans* as distinct from *Thielaviopsis paradoxa*, but Höhnel (1909) treated it as synonymous with *T. paradoxa.* Our attempts to locate the type specimen of *E. luteolum* have been unsuccessful. Delacroix's collections were thought to be in the Muséum National d'Histoire Naturelle, Paris, but Mme. Nicot could find no specimen there. *E. luteolum* must be considered a *nomen dubium*, at least until the fate of the holotype is known.

Höhnel (1925) described *Endoconidium abietinum* as follows: "Fructifications pulvinate or spherical, often coalescing, 100-300 μ high and wide, black. Basal layers hemispherical, composed of brown pseudoparenchymatous cells 2-3 μ wide, subtended by densely packed, much branched, broom-like clusters of conidiophores 30-35 μ long. The fertile branches [phialides?] are rather crooked, brown, 1.6-2.5 μ wide, sparsely septate below, open above; the hyaline, unicellular, long- or short-cylindrical conidia, 2.3 x 1.5 μ, are formed in short rows in the interior of the conidiophores, and discharged through the open ends. On the under side of fallen conifer needles at Sonntagsberg, Lower Austria, May 1913, P. P. Strasser" [freely translated from the German text]. The Farlow herbarium houses a specimen consisting of a single slide. The label on the folder reads "n.3178 Herb. Prof. Dr. Fr. v.Höhnel. *Endoconidium abietinum* v.H. n.sp. unterseite inner Tannennädeln." The slide shows the conidiogenous cells to be phialides arranged in dense fascicles (Figure 9), ampulliform, yellowish brown to pale brown, 9-16 μ long, composed of a subcylindrical venter 5.5-10 x 3.0-4.0 μ and an obconical collarette 3.5-6.0 μ deep and 2.0-3.0 μ wide. The conidia are enteroblastic-phialidic, ovoid to ellipsoidal, unicellular, hyaline, 3.0-5.5 x 1.0-2.0 μ. The morphology of the phialides and the conidia places the fungus in *Phialophora* Medlar, but although Cole and Kendrick (1973) have published a partial account of the genus, we still cannot place this fungus satisfactorily.

Columnophora (Figure 10)

Bubák and Vleugel (apud Bubák 1916) transferred *Oospora rhytismatis* Bresad. to their new genus *Columnophora* Bub. & Vl., as *Columnophora rhytismatis* (Bres.) Bub. & Vl., with the following generic description: "Parasitica, erumpens. Conidia cuboideo-globosa, ovoideis v. ellipsoideis, olivacea, glabra, 3-4-catenulata *ex interiore conidiophororum exsilientia.* . . . Genus Sporoschismati v. Thielaviopsi forte affinis" [italics ours].

The salient features of the species were as follows: "Caespitulis densiusculis, griseo-chlorinis, erumpentibus, pulvinatis, tomentosis; conidiophoris tereti-oblongis, continuis v. 1-2-septatis, 25-45 x 10-13, saepe inflatis, olivaceo-brunneis; conidiis 3-4-catenulatis concoloribus, levibus, 13-25 x 9-13, utrinque subtruncatis, conidio apicali supra tenuato-rotundatis, submitriformi." The Bresadola herbarium in S houses a specimen, S # 300, labelled "*Oospora rhytismatis* Bresadola ex Herb.

Diagrammatic interpretation of development of phialoconidia and chlamydospores in the genera studied. I and II-Enteroblastic-phialidic development of conidia and chlamydospores; III-holoblastic development of chlamydospores; IV-thallic development of chlamydospores.

Bresadola auf der unterseite . . . im Lechengrunde bei Obernwiesenthal (Erzgebirge), 29 Sept. 1903." In the same herbarium is another specimen of the fungus labelled: "Krieger, Fungi Saxonici # 2345 *Oospora rhytismatis* Bres. nov. sp . . . auf der unterseite von *Rhytisma salicinum* auf *Salix aurita* im Lechengrunde bei Obernwiesenthal, nur einmal gefunden, 29 Sept. 1903, leg. W. Krieger." Through the courtesy of Dr. Sten Ahlner, we have been able to examine both specimens. The fungus forms erumpent stromata on the lower surface of the leaves infected by *Rhytisma*. The conidiophores form a compact palisade and are cylindrical to subcylindrical, 0-2-septate in the basal part, subhyaline to pale brown, and terminate in an annellide. The conidia are holoblastic, cylindrical or ellipsoidal, 2-septate, subhyaline to pale brown, 19-41 x 8-12 μ, walls thick and verrucose (rarely smooth). Mature conidia have a tendency to break up at the septa.

This fungus appears to fall within the generic limits of *Stigmina* Sacc., but its affinities at the species level are yet to be determined.

9) *Phialophora* sp. from type of *Endoconidium abietinum*.

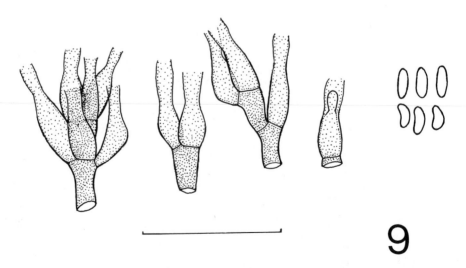

9

0)   *Stigmina* sp. ex type of *Columnophora rhytismatis.*

10

Discussion

In the past, the genus *Chalara* has been distinguished from other genera by Sac-cardoan criteria. Cylindrical, hyaline conidia extruded in long chains from the interior of mononematous conidiophores were considered as typical of the genus. *Thielaviopsis* and *Chalaropsis* were segregated from *Chalara* because they possessed additional propagules—thick-walled and darkly pigmented. The recognition by Hughes (1953) that the conidiogenous cell in *Chalara insignis* is a phialide subtly changed the concept of the genus, shifting the emphasis of the generic criteria from the conidia to the conidiogenous cell. The chief diagnostic characters of *Chalara* and allied genera thus lie in the phialides, which invariably have a well-differenti-ated venter and a long collarette. The transition from venter to collarette can be abrupt, gradual or almost imperceptible. The walls of the conidiophores (phialo-phores) or the phialides are generally relatively thick at the base and become progressively thinner towards the apex. The conidia are enteroblastic-phialidic (phi-aloconidia) and are released by circumscissile dehiscence of the thin apical wall of the phialide. Accessory criteria like the shape and size of the conidia, whether they have obtuse, rounded or truncate ends, the presence or absence of a basal marginal frill, septation and pigmentation, help to differentiate the species.

The dimorphic genera *Thielaviopsis, Chalaropsis,* and *Stilbochalara*, unlike the monomorphic *Chalara*, have presented some problems of taxonomy and nomencla-ture. In the taxonomic and nomenclatural treatment of pleomorphic imperfect fungi much stress is laid on the interpretation of Article 59 of the International Code of Botanical Nomenclature. Many conidial fungi have no known sexual states, and many others, whose relationships with sexual states have been established, often occur in nature separately, or fail to produce the sexual states when grown in culture. Recognizing that such imperfect fungi need names, the Code allows provi-sion of binomials for them even if they have known sexual states. In referring to imperfect fungi, however, the Code does not attempt to distinguish between those having only one conidial state (monomorphic) and those having two or more conid-ial forms (pleomorphic) developing simultaneously or successively, in close associa-tion or otherwise. Following Mason (1937), Hennebert (1971) has drawn our atten-tion to two possible interpretations of Article 59 of the Code:

1) In a literal interpretation, there will be a single binomial for a monomorphic imperfect fungus, and, for a pleomorphic imperfect fungus, as many binomials as the states that are recognized. The effect of this interpretation is that each particular state of a pleomorphic imperfect fungus has a binomial, but the species as a whole is left without a proper botanical name. Hennebert (1971) wrote: "If then it was desired to name the composite polymorphic species without amending one of the available state names, the only procedure would

be to give it a new specific epithet to accommodate that pleomorphy. This method, seen from a Linnaean point of view, would be inadmissible under the Code, because such an epithet would automatically be superfluous if its type included the type of an earlier name or synonyms of these earlier names."

2) The second interpretation is implied in the principle: "Only one binomial is admissible for each species." According to Hennebert (1971) such an interpretation would allow synonymy between perfect and imperfect names. In the case of the pleomorphic imperfects, since all epithets applied are nomenclaturally equal, priority will dictate the choice of the oldest epithet, and its type, no matter what its morphology, will typify the composite species, all other epithets being taxonomic synonyms.

Extending these interpretations, Hennebert (1971) has analyzed three possible approaches to the classification of pleomorphic imperfect fungi:

1) *Anatomical system:* The genus is a form-genus, and every species a form-species. The type of the name of a form-genus is a form-species whose name is fixed to a single, morphologically distinguishable, imperfect state. A pleomorphic imperfect fungus is assigned as many binomials in different genera as it has distinguishable imperfect states, no one name being correct for the whole fungus. Each form-genus is necessarily monomorphic and synonymy between the names of form-genera is restricted to those based on morphologically comparable single states; e.g., *Echinobotryum atrum* Corda for the chlamydospores, and *Cephalotrichum stemonitis* (Pers.) Link ex Fr. for the holoblastic-annellidic conidia of one fungus, neither specific nor generic names going into synonymy.

2) *Botanical system:* The genus is an imperfect genus, and every species a whole imperfect species. The type of the name of an imperfect genus is an imperfect species whose name is fixed to the whole fungus. Only one name is correct for a pleomorphic fungus. Imperfect genera may be monomorphic or pleomorphic, with strict priority applied to the synonymy of their names; e.g., *Cephalotrichum stemonitis* = *Echinobotryum atrum,* and, because this species is the type of both generic names, *Cephalotrichum* Link ex Fr. = *Echinobotryum* Corda.

3) *Botanico-anatomical system:* The genus is a form-genus, but each species is a whole imperfect species. The type of the name of the form-genus is an imperfect species but not as a whole, only in its sole state (if it is monomorphic) or one selected state—the lectostate—if it is pleomorphic. Binomials are not assigned to states but only to entire imperfect species. Synonymy between specific names occurs when connections between states are discovered, with strict priority applied to epithets; but synonymy of generic names is restricted as in the anatomical system, because the form-genera remain of monomorphic value by the typification of their names. The choice of generic name is a matter of taxonomic judgment; e.g., *Cephalotrichum stemonitis* st. *Cephalotrichum* (for the annellidic state), *C. stemonitis* st. *Echinobotryum* (for the chlamydosporic state) syn. = *E. atrum.*

This is a most difficult and complex problem, which has not yet received the attention it deserves. Hennebert's (1971) contribution provides a base for the extensive discussions which must take place before a consensus can be achieved. Initially *Chalara* is monomorphic and presents no problems. As soon as we decide to include the species of *Chalaropsis* and *Thielaviopsis* in *Chalara,* we are faced with the problems of pleomorphy. We could apparently adopt either a Botanical (2) or a Botanico-Anatomical (3) system. One weakness of type (3) is that it is quite possible for one author to choose one conidial state as lectostate, while another author chooses a different conidial state (compare our treatment of *Chalaropsis* with that of Hennebert [1967]). However, if we are to be able to include the diversity represented by the species of *Chalara, Chalaropsis* and *Thielaviopsis* in a single genus, we must choose that state which is present in all the species under considera-tion. This is unequivocally the phialidic or *Chalara* state, which therefore becomes the lectostate. This is essentially the same approach as that advocated by Hughes (1953) and Barron (1968). Hughes wrote: "In such a pleomorphic taxon, therefore, the generic name is the result of classifying one of the states or spore forms to the exclusion of all others. Mason advocated the use of differential names for classify-ing spore forms (states) and this seems to be the only way out with the Fungi Imperfecti. The subsidiary states may be referred to as *Hyalodendron* blastospores, *Cephalosporium* phialospores, etc., as the case may be, without the unnecessary new combinations. If no suitable qualifying generic name is available precisely to indicate the type of subsidiary state then this can be referred to by the use of the spore terminology alone; but for this to work, of course, it is necessary to establish types of conidium development on a firm basis." This approach has recently been endorsed and extended by the participants of the First International Conference on Criteria and Terminology in Fungi Imperfecti, held at Kananaskis, Alberta (see Kendrick 1971). We have used spore terminology to describe subsidiary states, where these are present.

Went's (1893) circumscription of *Thielaviopsis* ". . . Conidia dimorpha, majora catenulata, ovata, fusca; minora cylindracea, hyalina, ex interiore hypharum catenu-latim generata et ex apice exsilientia" implies that the presence of both conidial states is the essential generic criterion. A similar criterion was established by Pey-ronel (1916) for *Chalaropsis*: "Conidia dimorpha; majora subglobosa vel subellip-tica, fusca, unicellularia, sessilia vel in brevibus mycelii ramis acrogena; minora cylindracea, hyalina, catenulata ex apice eorum exsilientibus; est *Thielaviopsis* macroconidiis simplicibus, non catenulatis." Hennebert (1967) emphasized ". . . the difficulties one would encounter in the nomenclature of the fungi imperfecti if the association of spore forms served as a diagnostic basis of the genera, or in other words, if form-genus names were based on more than one spore form. I believe that names of form genera, to be even more useful in designating definite conidial states in perfect fungi and in polymorphic imperfect fungi, should be un[equ]ivocal and therefore monomorphic. . . . Therefore, I consider the 'aleuriospore' state like that described in *Chalaropsis punctulata* to be the base of *Chalaropsis* Peyronel and propose amending the concept of the genus and restricting it to that state." This taxonomic treatment leads to some difficulties in that the fungus in its entirety is without a name, and since a name for the phialidic state inevitably becomes neces-sary, there will be a multiplicity of 'state-generic' names. We also consider Henne-bert's choice of state unfortunate, because the thick-walled, dark propagules of *Chalaropsis* are, in our view, less characteristic than the phialides. Many other hyphomycetes produce thick-walled, darkly pigmented propagules similar to those of *Chalaropsis*. By themselves, the thick-walled conidia of *Thielaviopsis paradoxa* can pass for those of *Mammaria* Ces. They might also be confused with *Acremo-niula, Humicola, Allescheriella* and possibly others (see Kendrick & Carmichael 1973, Plates 19, 21). There has been an instance of a species of *Scytalidium* Pesante being determined as an unknown species of *Thielaviopsis* because it produced chains of thick-walled, dark brown conidia. The catenulate or non-catenulate char-acter of the thick-walled, coloured conidia does not afford a firm basis for separat-ing *Thielaviopsis* from *Chalaropsis* since, as our ontogenetic studies established, both conditions can exist in the same fungus (e.g., *T. paradoxa*) at different stages of its growth.

Our studies have revealed interesting features which argue strongly for consideration of *Thielaviopsis* and *Chalaropsis* under *Chalara*. The conidiogenous cells (phialides) in members of the first two genera have all the essential features of those found in *Chalara*, with minor differences in pigmentation and wall thickness. The conidia are enteroblastic-phialidic (phialoconidia), cylindrical, unicellular, and hyaline at first, but often become darkly pigmented and thick-walled with age. Our ontogenetic studies of these thick-walled conidia have revealed that in *Thielaviopsis paradoxa* they are predominantly enteroblastic-phialidic, but some are holoblastic, occurring in chains or clusters, or singly. These enteroblastic-phialidic and holoblastic thick-walled conidia are morphologically indistinguishable. Electron microscopy may reveal differences in wall layers between the two types of conidia. In *Chalaropsis thielavioides*, and in the *Chalaropsis* state of *Ceratocystis radicicola*, the thick-walled conidia are predominantly holoblastic. In *Thielaviopsis basicola*, however, the thick-walled conidia are thallic-arthric. Whatever their mode of development, they are always produced at a later stage than the phialoconidia, and the fungus even at this phase could revert to the production of phialoconidia if favourable conditions were reestablished. These features, together with attributes like thick walls and dark pigmention, strongly suggest that these conidia essentially function as 'chlamydospores.' Features similar to those observed in *Thielaviopsis paradoxa* (conidial *Ceratocystis paradoxa*) are also found in the conidial states of *Ceratocystis adiposa* and *C. fimbriata*. De Seynes (1886), Went (1893), Ferraris (1910) and Peyronel (1916) employed the term 'macroconidia' to designate the thick-walled, dark brown conidia occurring in these taxa. The term macroconidia was used to indicate conidia of a larger diameter than the conidiophores, and we consider it ambiguous. Hennebert (1967) used the term 'aleuriospore' to designate the holoblastic conidia of *Chalaropsis*. 'Aleuriospore' was originally coined by Vuillemin (1911) for propagules intermediate between the chlamydospore and the conidium. He stated that these occurred on undifferentiated hyphae either at the tip, along the sides, or in an intercalary position and were released by the disintegration of part, or all, of the spore-bearing hyphae. He concluded that they "différent des conidies par leur union dissoluble avec les filaments myceliens." The participants in the First International Conference on Criteria and Terminology in Fungi Imperfecti (1969—see Kendrick 1971) rejected 'aleuriospore' as a confused term, though it is tempting to retain some derivative of the word to refer to the highly characteristic mode of dehiscence which Vuillemin was describing when he created the term. The term 'chlamydospore' has generally been used to denote an intercalary, indehiscent, thick-walled propagule. But this term as accepted at the Kananaskis conference is more inclusive, and embraces thick-walled spores, often with thallic development, whether terminal, intercalary, or on a short lateral stalk—in fact, most of the forms that were previously called aleuriospores as well as chlamydospores. Since the terms macroconidia and aleuriospores are no longer available for the thick-walled pigmented conidia of *Chalaropsis* and *Thielaviopsis*, and since we have demonstrated that these propagules are both enteroblastic-phialidic and holoblastic in at least three of the dimorphic taxa studied, the term 'chlamydospore,' even more broadly defined than at Kananaskis, is a convenient and appropriate catchall for them.

The occurrence, in all species of *Thielaviopsis* and *Chalaropsis* and in some *Ceratocystis* species, of phialides with overall similarity to those of *Chalara* strongly suggests that the names for the phialidic states of these taxa should be sought in *Chalara*. In this treatment, the state assignable to *Chalara* is considered to be the principal and most characteristic conidial state; the thick-walled, darkly pigmented chlamydospores are considered as a subsidiary state. In the case of *Thielaviopsis paradoxa*, the name *Chalara paradoxa* (de Seynes) Sacc. provides an earlier epithet. Tsao and Bricker (1970) argued for segregation of *Thielaviopsis basicola* from *Thielaviopsis* in view of the differences between it and the type species in the ontogenies of their thick-walled conidia. We agree, in part, with this view, but consider that the most natural nomenclatural and taxonomical niche for the fungus is in *Chalara* rather than a new genus. New names in *Chalara* have been provided for *Thielaviopsis basicola* and *Chalaropsis thielavioides*. In cultural behaviour and in morphology and dimensions of phialides and conidia, *Hughesiella euricoi* is identical with *Chalara paradoxa*, of which it is here considered a synonym. The synnematous form of *Stilbochalara dimorpha*, the type species of the monotypic genus *Stilbochalara,* shows a strong resemblance to *Chalara paradoxa*,which is also known to form occasional lax coremia. In the morphology and ontogeny of both kinds of conidia, these two taxa are identical. The fact that the holotype of *S. dimorpha* bears perithecia of a *Ceratocystis* species assignable to *Ceratocystis paradoxa*, the sexual state associated with *Chalara paradoxa*, supports the reduction of *S. dimorpha* to synonymy with *Chalara paradoxa.*

Excioconidium cibotti, the type species of *Excioconidium*, produces enteroblastic-phialidic conidia from phialides characteristic of *Chalara*. Plunkett (apud Stevens 1925) differentiated *Excioconidium* from *Chalara* on the basis of the septation of its conidia, a character we do not in this case regard as significant at the generic level. *Excioconidium* is here treated as a synonym of *Chalara*.

Fusichalara, though resembling *Chalara* superficially, has been considered distinct in view of the convex thickening of the inner wall of the phialide at the point of transition from venter to collarette, and the formation of two kinds of conidia from the same phialide with the sigmoid or fusiform, coloured or hyaline conidia predominating. We accept these differences as valid generic criteria.

Chaetochalara and *Sporoschisma* have phialides similar to those of *Chalara,* but have been regarded as distinct genera by virtue of the characteristic sterile elements consistently associated with their phialides (setae in *Chaetochalara*,capitate hyphae in *Sporoschisma*). Whether setae or capitate hyphae should be considered generic criteria may be questioned, particularly since we have already lumped under *Chalara* the species of *Chalaropsis* and *Thielaviopsis*, which produce chlamydospores in addition to *Chalara*-type phialoconidia. However, it must be clearly understood that we placed these generic names in synonymy with *Chalara* in order to resolve conflicts produced by the original authors' obligate fusion under a single generic name of disparate elements, each of which is, or could be, used to characterize a generic name by itself. No such conflict arises in the cases of *Chaetochalara* and *Sporoschisma*. Here the existing generic delimitation from *Chalara* is based on the consistent presence of specific and characteristic sterile elements. It may be suggested that we are segregating genera on the basis of a single character—presence or absence of setae or capitate hyphae as the case may be. We submit, however, that setae or capitate hyphae are character complexes, rather than single characters. Were we to 'score' or 'analyze' these structures for numerical taxonomy, we would be able to dissect them into many 'unit characters'—configuration, length, width, septation, pigmentation, wall thickness, ornamentation, etc. Since *Chalara* is now a large genus, and since *Chaetochalara* and *Sporoschisma* already exist, and can be easily keyed out (see page 59), we have maintained them both, at least for the present.

Another and potentially more important feature is the association of sexual and asexual states. Unfortunately, our knowledge of this aspect is still fragmentary. While most species of *Chaetochalara* have not been found with a sexual state, specimens of two species of *Chaetochalara* are intimately associated with apothecia of unnamed species of *Hyaloscypha* and *Calycellina*. This association may connote a genetic relationship, especially since it has been found in specimens from such disparate locations as North America and New Zealand. Full descriptions of these discomycetes are given on page 183. The known sexual states of two species of *Sporoschisma* are classified in *Melanochaeta,*the perithecia of which bear capitate hyphae, and, occasionally, *Sporoschisma* phialides. Finally, genetic relationships between some *Ceratocystis* species and their *Chalara, Chalaropsis,* or *Thielaviopsis* states have been established beyond any doubt. In our opinion, these facts offer indirect evidence in support of our separation of *Chalara, Chaetochalara* and *Sporoschisma*.

Ascoconidium is distinguished from other related genera by its clavate, thick-walled conidiogenous cell (phialide), which lacks a well-differentiated venter, has the conidiogenous locus near its base, and releases the conidia by a vertical split in, rather than a circumscissile dehiscence of, the apical wall. *A. tsugae* has recently been connected with its sexual state, *Sageria tsugae*, a helotiacious discomycete.

The conidiophores (phialophores) in *Bloxamia* and *Endosporostilbe* are densely aggregated into sporodochia or synnemata, and the phialides are subcylindrical without a well-differentiated venter. These features segregate them from other genera considered here. Illman (1964) considered *Endosporostilbe nilagirica* as congeneric with *Stilbochalara dimorpha (=Chalara paradoxa)*, but in *E. nilagirica*, the type and only species of *Endosporostilbe*, the phialides differ from those of a *Chalara*, and the thick-walled, darkly pigmented propagules so characteristic of *Chalara paradoxa* are entirely absent. *E. nilagirica* may be disposed in *Bloxamia*, and *Endosporostilbe* is here regarded as a synonym of *Bloxamia*.

In passing, we wish to offer some conjectures on the possible evolutionary significance of some of the features found in these genera, and the relationships they suggest. Savile (1968) has put forward a hypothesis that the fungi evolved from a parasitic to a saprophytic mode of life. He stated: "Obligate parasitism in the fungi is not a belated evolutionary bypath, but a fundamental attribute of primitive groups; and saprophytism arose repeatedly from it. The first typical fungi were probably parasites; and it was as parasites that the fungi left the water protected by the tissues of their hosts." If this hypothesis is accepted, then most of the dimorphic *Chalara* species may have originated from a parasitic ancestral form. The *Chalara* state of *Ceratocystis adiposa*, the *Chalara* state of *Ceratocystis fimbriata, Chalara paradoxa, 'Chalaropsis' thielavioides,* and *'Thielaviopsis basicola'* are known to be parasitic on various higher plants. Possibly because of the restricted life span of the host, and perhaps because of the hyaline, thin-walled nature of their vegetative hyphae, conidiophores and conidia, selective pressures might favour those fungi which produce a long-term survival mechanism such as thick-walled, darkly pigmented chlamydospores. In *Chalara paradoxa*, this mechanism manifests itself indiscriminately in the hyphae, conidiophores and conidia. This may partially explain the apparently aberrant mixing of enteroblastic-phialidic and holoblastic modes of development in such propagules. In more advanced forms these mechanisms may be built into specific structures, viz., enteroblastic-phialidic and holoblastic conidia (which may both be termed 'chlamydospores'), or sclerotia, e.g., *Chalara thielavioides*, the *Chalara* state of *Ceratocystis radicicola*, and *Chalara quercina*. In what might be considered even more advanced forms, the survival mechanism may be represented by pigmentation and thickening of the wall of the conidiophores themselves. From this level, divergent evolution may have led to clustering of conidiophores into lax coremia or into synnemata or sporodochia or to the development of thick, pigmented walls in both the conidiophores and conidia. Parallel with these postulated lines of evolution, production of a large number of small unicellular conidia might be replaced or supplemented by the formation of larger conidia with more than one cell. Using these conjectures as a starting point we may suggest that *Sporoschisma, Fusichalara, Ascoconidium* and *Bloxamia* represent more highly evolved forms than the dimorphic species of *Chalara*. The significance of sterile structures like setae and capitate hyphae is not yet understood, though they may serve to protect developing conidiophores from grazing arthropods.

In conclusion, this taxonomic treatise, in which the genera are differentiated on the basis of the phialidic state, suggests an arrangement of the different taxa based on what might be their natural relationships, and also eliminates some superfluous generic and species names, in conformity with the avowed objectives of taxonomy—simplicity, practicality and convenience. An explanation has been suggested for the existence of two types of conidia seen in the old dimorphic genera *Thielaviopsis* and *Chalaropsis*, and we consider it both logical and practical to unite these genera under the single generic name *Chalara*, providing an emended generic concept to include these taxa. Perhaps our taxonomic conclusions may be summed up most succinctly by the key to genera in the following pages.

Taxonomic Part

Classification:
Kingdom : Fungi
Division : Eumycota
Subdivision : 'Deuteromycotina' (Fungi Imperfecti)
Form Class : Moniliales
Ontogenetic group : blastic
Subgroup : enteroblastic-phialidic

Conidia enteroblastic-phialidic, produced in basipetal succession from a fixed, in most cases deep-seated, conidiogenous locus within a usually determinate, stable conidiogenous cell (phialide); phialides usually pigmented, and possessing a *deep, cylindrical* or *subcylindrical collarette* which often determines the shape of the conidia delimited within it.

Key to Genera

1) Phialophores aggregated into sporodochia or synnemata; conidia
   cuboid . . . .**Bloxamia** (page 167)
1) Not as above 2

2) Phialophores consistently associated with sterile setae or capitate hyphae . . .3
2) Phialophores not associated with such sterile elements 5

3) Sterile elements in the form of pigmented setae without apical mucilaginous (?)
   caps . . . .**Chaetochalara** (page 148)
3) Sterile elements in the form of capitate hyphae with apical, hyaline, mucilagi-
   nous (?) caps 4

4) Capitate hyphae resembling young phialophores; phialophores characteristically with long,percurrent proliferations; collarette shallow or cupulate, usually wider than venter; ameroconidia dark, angular or wedge-shaped, widest
   at apex . . . . **Catenularia**
4) Capitate hyphae much narrower and shorter than phialophores; collarette
   very deep, cylindrical; phragmoconidia dark,
   cylindrical . . . . **Sporoschisma** (page 157)

5) Phialides clavate, lacking a morphologically differentiated venter, wall uniformly
   thick, splitting vertically at apex of collarette to release conidia; phragmo- or
   dictyo-conidia hyaline, cylindrical . . . .**Ascoconidium** (page 164)
5) Phialides not clavate, with differentiated venter, wall attenuated toward collarette apex which is lost by circumscissile split to release conidia 6

6) Conidia accumulating in a slimy droplet at apex of collarettes 7
6) Not as above 8

7) Phialophores with well-defined stipe, bearing a complex apical penicillate conidiogenous apparatus, the ultimate tier of metulae subtending numerous
phialides . . . . **Phialocephala**
7) Phialophores much simpler; stipe often lacking, metulae absent; phialides solitary
or in small clusters . . . . **Phialophora**

8) Phialides borne singly at apex of phialophores, or sessile 9
8) Phialides borne in simple radiating clusters at apex of
phialophores . . . . **Sporendocladia** (page 162)

9) Collarette shallow, cupulate, or funnel shaped, usually wider than
venter . . . .**Sporoschismopsis**
9) Collarette deep, cylindrical, narrowly conical or obconical 10

10) Phialides with a convex thickening of the inner wall at the zone of transition
from venter to collarette; dimorphic conidia produced from each phialide;
first conidium cylindrical, subsequent conidia fusiform or
sigmoid . . . . **Fusichalara** (page 144)
10) Phialides without localized thickening of the inner walls; conidia monomorphic, mostly cylindrical; chlamydospores present in some
species . . . .**Chalara** (page 60)

In the species descriptions that appear on the following pages, records of lengths of
conidiophores *include* the lengths of phialides. The general mean of a quantitative
character is given in square brackets, e.g. 5-7[6.5]μ. Where they occur, extreme
measurements are indicated in round brackets by numerals accompanied by a
hyphen, e.g. 5-7(-9)μ. Type specimens are clearly indicated under 'specimens examined.'

In the key to species, numerals appearing in parentheses after species' names
indicate the order in which the descriptions appear. To facilitate identification, the
illustrations immediately follow the key. As far as possible illustrations have been
prepared to the same scale. Unless otherwise stated, the scale on an illustration
represents 20μ.

Chalara (Corda) Rabenh.
in Kryptogamenflora 1: 38, 1844; description emended Nag Raj and Kendrick.
≡ *Torula* (Pers.) Link subg. *Chalara* Corda in Icones Fung. 2:9, 1838.
≡ *Torula* (Pers.) Link sect. *Chalara* (Corda) Corda in Icones Fung. 5:5, 1842.
    = *Cylindrocephalum* Bon. in Handb. Allg. Mykol., 1851 fide Hughes in Can.
        J. Bot., 36: 747, 1958.
    = *Thielaviopsis* Went in Arch. voor de Java Suekerr., 1893, p. 4.
    = *Stilbochalara* Ferd. & Winge in Bot. Tidsskr., 30: 220, 1910.
    = *Chalaropsis* Peyr. in Staz. sper. agr. ital., 49: 595, 1916.
    = *Excioconidium* Plunkett apud Stevens in Bull. Bernice P. Bishop Mus., 19:
        156, 1925.
    = *Hughesiella* Bat. & Vital in Anais Soc. Biol. Pernamb., 14: 142, 1956.

Phialophores simple, rarely branched, scattered or aggregated, sometimes reduced to
phialides, or cylindrical, septate, hyaline, subhyaline, brown to dark brown, terminating in a phialide. *Phialides* ampulliform, lageniform, obclavate, urceolate or sub-

cylindrical, composed of a well-differentiated venter and usually a long collarette; transition from venter to collarette abrupt, gradual or barely perceptible. *Conidia* enteroblastic-phialidic, cylindrical, obclavate or ellipsoid, with rounded or truncate ends, often provided with basal marginal frill or rarely fringes of wall material,

) Reciprocal illustrations supplementing generic key [not drawn to scale].

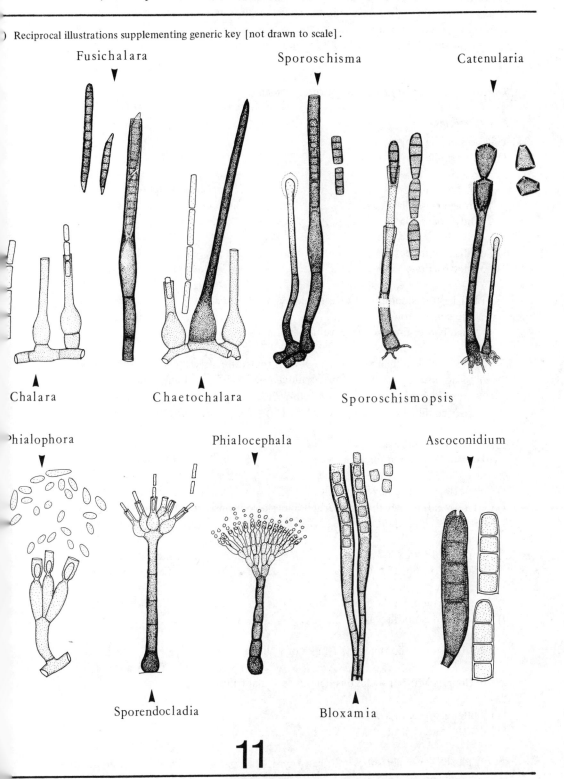

Fusichalara

Sporoschisma

Catenularia

Chalara

Chaetochalara

Sporoschismopsis

Phialophora

Phialocephala

Ascoconidium

Sporendocladia

Bloxamia

11

unicellular or septate, mostly hyaline, less frequently subhyaline or pale brown, formed singly or in chains. *Chlamydospores*, when present, are either converted phialoconidia or thallic-arthric, or holoblastic, short-cylindrical, doliiform, ovoid, pyriform, ellipsoid, globose, subglobose or irregular; unicellular or in some species septate; subhyaline, brown to dark brown; walls predominantly thick, smooth, verrucose, ridged, papillate or fimbriate, with or without vertical, oblique or transverse germ slits; intercalary or solitary and terminal, or in terminal chains or clusters.

Type species: Chalara fusidioides (Corda) Rabenh.

Key to species

1) Phialoconidia septate 2
1) Phialoconidia non-septate 32

    2) Dictyoconidia .... **Chalara dictyoseptata** (23)
    2) Phragmoconidia 3
    2) Didymoconidia 9

3) Conidia 3-septate 4
3) Conidia 7-septate 6

    4) Phialophores composed of a single stalk cell and a phialide; venter and collarette concolorous .... **Chalara unicolor** (60)
    4) Phialophores multi-septate; venter and collarette not concolorous 5

    5) Collarette darker than venter, rough with transverse striae; phialides subcylindrical to lageniform, 105-145 μ long; conidia 22-37 x 4-5 μ .. **Chalara inflatipes** (34)
    5) Collarette lighter than venter, smooth-walled; phialides obclavate to cylindrical, 13-40 μ; conidia 8-18 x 2-3 μ, occasionally 0-1-2-septate... **Chalara pteridina** (46)

    6) Conidia cylindrical 7
    6) Conidia clavate or obclavate .... **Chalara cibotti** (16)

7) Collarette darker than venter 8
7) Collarette not darker than venter; phialides pale brown; conidia 39-56 x 8.5-11 μ .... **Chalara pulchra** (47)

    8) Conidia 18-54 x 5-7 μ .... **Chalara insignis** (35)
    8) Conidia 50-66 x 5.5-6 μ .... **Chalara bicolor** (9)

9) Conidia verrucose at both ends .... **Chalara cladii** (17)
9) Conidia with smooth walls 10

    10) Conidia with conspicuous fringes of wall material at the base or at each end .... **Chalara rubi** (51)
    10) Conidia without such conspicuous terminal fringes 11

11) Venter asperate 12
11) Venter smooth 14

    12) Phialides usually subcylindrical; transition from venter to collarette gradual; conidia 15-19 x 2.5-3 μ .... **Chalara scabrida** (52)

12) Phialides usually ampulliform; transition from venter to collarette
    abrupt **13**

13) Ratio of mean lengths of collarette/venter = 1.9:1; mean conidium l/w ratio =
    4.5:1; conidia without basal marginal frills . . . . **Chalara emodensis** (26)
13) Radio of mean lengths of collarette/venter = 5.3:1; mean conidium l/w ratio =
    10:1; conidia bearing basal marginal frill . . . .**Chalara curvata** (20)

14) Collarette darker than venter **15**
14) Collarette not so **18**

15) Collarette uniformly dark **16**
15) Collarette darker in its basal part, but lighter above **17**

16) Conidium unequally 2-celled; phialophores
    pluri-septate . . . .**Chalara inaequalis** (33)
16) Conidial septum median; phialides usually sessile . . . . **Chalara nigricollis** (39)

17) Phialides urceolate; phialophores few- to many-septate; transition from venter
    to collarette abrupt . . . . **Chalara tubifera** (58)
17) Phialides subcylindrical, rarely lageniform; phialophores usually reduced to
    phialides or a phialide and a stalk cell; transition from venter to collarette
    gradual . . . . **Chalara agathidis** (3)

18) Conidia bearing marginal frill at the base **19**
18) Conidia without basal frill **27**

19) Phialophores arising from stromatic aggregations of hyphal cells **20**
19) Phialophores not arising from stromatic aggregations **26**

20) Phialophores reduced to a stalk cell and a phialide **21**
20) Phialophores well developed and many-septate **23**

21) Phialides subcylindrical; collarette width 2.5 μ or less; conidia 1.5-2 μ wide;
    vegetative hyphae asperate . . . . **Chalara gracilis** (31)
21) Phialides not subcylindrical; collarette usually more than 2.5 μ wide; conidia
    generally more than 2 μ wide; vegetative hyphae smooth **22**

22) Conidia 8-13 μ long; mean conidium
    l/w ratio = 3.7:1 . . . .**Chalara kendrickii** (36)
22) Conidia 13-18.5 μ long; mean conidium
    l/w ratio = 4.6:1 . . . . **Chalara angionacea** (5)

23) Phialophores with repeated percurrent proliferations . . . .**Chalara prolifera** (45)
23) Phialophores not repeatedly proliferating **24**

24) Collarette widening toward apex; conidia 12-20 μ long . . **Chalara acuaria** (1)
24) Collarette not widening toward apex **25**

25) Collarette cylindrical; transition distinct; conidia 2.5-3 μ
    wide . . . .**Chalara rostrata** (50)
25) Collarette narrowing toward apex; transition indistinct; conidia 3.5-4 μ
    wide . . . . **Chalara selaginellae** (53)

26) Phialides 59-69 (-72) μ long; collarette 5-6.5 μ wide; conidia 3.5-5 μ
    wide . . . . **Chalara aotearoae** (6)
26) Phialides 30-34 μ long; collarette 2.5-3 μ wide; conidia 2-2.5 μ
    wide . . . . **Chalara stipitata** (56)

27) Phialophore reduced to phialide alone, or phialide and one or two stalk
    cells **28**
27) Phialophores with more than two stalk cells **29**

28) Venter globose; ratio of mean lengths of
    collarette/venter = 3:1 . . . . **Chalara rhynchophiala** (49)
28) Venter ellipsoidal or subcylindrical; ratio of mean lengths of collar-
    ette/venter = 1.7:1 . . . . **Chalara hughesii** (32)
29) Phialides urceolate . . . . **Chalara urceolata** (61)
29) Phialides not so **30**

30) Phialophores arising from a thin stromatic layer of cells; phialides obclavate
    or lageniform; transition from venter to collarette gradual; conidia 8-19 x
    2-2.5 μ . . . . **Chalara aurea** (7)
30) Phialophores not arising from stromatic aggregations **31**

31) Mean conidium l/w ratio = 7.4:1; conidia often forming long, helicoid
    chains . . . . **Chalara spiralis** (55)
31) Mean conidium l/w ratio = 3.9:1; conidia not in helicoid
    chains . . . . **Chalara ginkgonis** (30)

32) Chlamydospores absent **33**
32) Chlamydospores present **59**

33) Conidia ellipsoidal or clavate **34**
33) Conidia cylindrical **37**

34) Phialoconidia exceeding 10 μ long . . . . **Chalara breviclavata** (12)
34) Phialoconidia less than 10 μ long **35**

35) Collarette asperate; base of phialophore usually broadly conical and
    darker . . . .**Chalara brunnipes** (15)
35) Collarette smooth; transition usually marked by a constriction **36**

36) Phialides usually sessile; collarette 6-8 μ long, conidia not exceeding 4 μ
    long . . . .**Chalara** state of **Ceratocystis autographa** (63)
36) Phialophores 1-4-septate; collarette 7-11 μ long; conidia up to 8 μ
    long . . . . **Chalara constricta** (18)

37) Phialophores rough-walled **38**
37) Phialophores smooth **40**

38) Collarette widening toward apex, 7-9.5 μ long; transition usually marked by
    a constriction . . . . **Chalara ellisii** (25)
38) Collarette cylindrical, longer; transition lacking constriction **39**

39) Collarette 6-15 x 1.5-3 μ; venter 4.5-6 μ wide; conidia 3-9.5 x
    1-1.5 μ . . . . **Chalara cylindrica** (21)

39) Collarette 15-39 x 3-4.5 μ; venter 6-9.5 μ wide; conidia 9.5-13 x
    2-3 μ . . . . **Chalara bohemica** (10)

40) Phialophores reduced to sessile phialides,or phialides with one or two stalk
    cells **41**
40) Phialophores well-developed and multi-septate **49**

41) Phialophores and phialides
    hyaline . . . . **Chalara** state of **Cryptendoxyla hypophloia** (70)
41) Phialophores and phialides pigmented **42**

42) Collarettes usually shorter than venter **43**
42) Collarettes as long as,or longer than,venter **46**

43) Venter subcylindrical **44**
43) Venter conic or ellipsoidal **45**

44) Collarette 6-8 μ long; phialides lageniform, 12-21 μ
    long . . . .**Chalara austriaca** (8)
44) Collarette 6-17 μ long; phialides usually obclavate, 18-36 μ
    long . . . . **Chalara microspora** (38)

45) Venter usually ellipsoidal, 4-6.5 μ wide; collarette 3-10 x 2-3 μ; transition usu-
    ally abrupt and marked by a constriction; conidia 1.5-2.5 μ
    wide . . . . **Chalara ampullula** (4)
45) Venter usually conic, 3-4 μ wide; collarette 10-11 x 1.5 μ; transition gradual
    and without constriction; conidia 1-1.5 μ wide . . . . **Chalara sessilis** (54)

46) Venter usually globose; phialides less than 28 μ
    long . . . . **Chalara fusidioides** (28)
46) Venter cylindrical, subclindrical or ellipsoidal **47**

47) Collarette darker at base; conidia with basal marginal
    frills . . . . **Chalara brevipes** (13)
47) Collarette paler than,or concolorous with,venter; conidia without marginal
    frills **48**

48) Conidia up to 19 μ long; collarette 2-3 μ wide . . . .**Chalara affinis** (2)
48) Conidia not exceeding 8 μ long; collarette 2.5-4 μ
    wide . . . .**Chalara fungorum** (27)

49) Phialides subcylindrical; transition imperceptible **50**
49) Phialides lageniform or obclavate; transition gradual or abrupt **51**

50) Conidia 3.5-8 x 2-3.5 μ; phialides 25-34 μ long, venter 3-4.5 μ wide; collar-
    ette 2.5-4 μ wide . . . .**Chalara quercina** (48)
50) Conidia 5.5-11 x 3.5-4.5 μ; phialides 50-61 μ long; venter 4.5-7.5 μ
    wide; collarette 4.5-6.5 μ wide . . . . **Chalara ungeri** (59)

51) Venter longer than collarette **52**
51) Venter as long as,or shorter than,collarette **54**

52) Mean conidium l/w ratio = 1.5:1.; conidia 2-4 x

1.5-2 μ . . . . .Chalara brevispora (14)

52) Mean conidium l/w ratio = 4:1 or more **53**

53) Collarette less than 10 μ long; transition gradual; conidia lacking basal frill; mean conidium l/w ratio = 4:1 . . . . **Chalara longipes** (37)

53) Collarette 10-15 μ long; transition usually abrupt; conidia with basal marginal frill; mean conidium l/w ratio = 5:1 . . . . **Chalara novae-zelandiae** (41)

54) Phialophores arising from basal stromatic aggregations of vegetative hyphae . . . .**Chalara germanica** (29)

54) Phialophores not arising from such aggregations **55**

55) Mean conidium l/w ratio = 2.6:1 or less **56**

55) Mean conidium l/w ratio more than 2.6:1 . . . .**57**

56) Venter 6.5-9.5 μ wide; collarette 3.5-5 μ wide; conidia 2.5-4 μ wide . . . .**Chalara brachyspora** (11)

56) Venter 4-7 μ wide; collarette 1.5-2.5 μ wide; conidia 1.5-2.5 μ wide . . . . **Chalara parvispora** (44)

57) Conidia with basal marginal frill . . . . **Chalara nothofagi** (40)

57) Conidia lacking basal frill **58**

58) Venter up to 8 μ wide; transition abrupt; conidia 1.5-2.5 μ wide; mean conidium l/w ratio = 6:1 . . . . **Chalara cylindrosperma** (22)

58) Venter not exceeding 5.5 μ wide; transition gradual; conidia 1-1.5 μ wide; mean conidium l/w ratio = 4:1 . . . .**Chalara crassipes** (19)

59) Chlamydospores ornamented **60**

59) Chlamydospores smooth **61**

60) Phialides 20-40 μ long, often rough-walled; chlamydospores usually fimbriate, 11-23 μ wide; germ slits obscure . . . . **Chalara** state of **Ceratocystis adiposa** (62)

60) Phialides 48-95 μ long, smooth; chlamydospores verrucose or usually striate, 11-15 μ wide; germ slit vertical . . . . **Chalara** state of **Ceratocystis radicicola** (69)

61) Chlamydospores in very characteristic rectilinear series of up to 7 segments (resembling phragmoconidia), segments short-cylindrical; germ slits transverse and bipolar . . . .**Chalara elegans** (24)

61) Chlamydospores usually solitary and/or sympodial; germ slits obscure or vertical **62**

62) Germ slits on chlamydospores vertical; chlamydospores 9.5-25 x 7-15 μ . . . .**Chalara paradoxa** (43)

62) Germ slits obscure or absent **63**

63) Phialides 20-36 μ long; venter 2.5-4.5 μ wide; collarette 7.5-14 x 2-3 μ; chlamydospores (converted phialoconidia) doliiform or ovoid, moniliform . . . . **Chalara** state of **Ceratocystis moniliformis** (67)

63) Phialides larger **64**

64) Chlamydospores sympodially arranged, subglobose or pyriform, 7.5-14 x
   6-11 μ; wall smooth or usually minutely verrucose . . . . . **Chalara ovoidea** (42)
64) Chlamydospores not sympodially arranged, variable in shape **65**

65) Chlamydospores usually globose, up to 18 μ
   wide . . . .**Chalara thielavioides** (57)
65) Chlamydospores usually ellipsoidal or pyriform, not exceeding 13 μ
   wide . . . .**Chalara** state of **Ceratocystis fimbriata** (66)

1) Chalara acuaria Cooke & Ellis (Figure 22A)
   in Grevillea, 6: 68, 1878.
   Colony superficial, effuse, brown, hairy. Phialophores arising from a thin pseudoparenchymatous layer in the substratum, simple, more or less cylindrical, 2-4-sep-
tate, not constricted at the septa, brown, smooth-walled, 50-120 [74] μ long and
5.5-9 [7] μ wide at the base, and terminating in a phialide. Phialides obclavate, pale
brown, 33-77 [54] μ long, composed of an ellipsoidal venter 15-26 [20] x 5-11 [8]
μ, and a cylindrical or more or less obconical collarette 14-57 [37] x 2.5-5 [3.8] μ;
transition from venter to collarette gradual; ratio of mean lengths of collarette and
venter = 1.8:1. Phialoconidia found singly or in short chains; cylindrical with a
rounded apex and a truncate base with minute marginal frill; 0-1-septate, hyaline
smooth-walled, 12-20 [16] x 2-3.5 [2.7] μ; mean conidium length/width ratio =
6:1.
Habitat: On needles of *Abies, Pinus, Podocarpus dacrydioides.*
Specimens examined: 1) M. C. Cooke 1885—Fungi of New Jersey. 2786, on fir
leaves, Newfield, N.J. U.S.A., J. B. Ellis (in IMI 112288 ex *Holotype* in Herb.
M. C. Cooke in K); 2) Ellis and Everhart, N. Amer. Fungi, 2nd Ser. # 2966, on
fallen leaves of *Pinus*, in FH, B and BPI; 3) PDD 32874, on *Podocarpus dacrydioides*, Big Tree Tr., Peel Forest Pk., Centerbury Prov., N.Z., 26.IV.1974,
B. Kendrick (KNZ 682a).

Known distribution: New Zealand, U.S.A.
   C. acuaria closely resembles *C. rubi* but differs in its more robust phialophores, the
shape and size of its phialides, its more or less obconical collarettes, and because its
conidia do not bear frayed fringes of wall material at one or both ends. It differs
from *C. rostrata* in producing obconic rather than cylindrical collarettes, brown and
few-septate, rather than usually dark brown and pluri-septate phialophores, and
longer conidia.

2) Chalara affinis Sacc. & Berl. (Figure 31B)
   in Atti Ist. veneto Sci., 3: 741, 1885.
   Colony superficial, effuse, cream yellow, crustose due to sinking in of the mass
of phialoconidia. Phialophores simple, erect, cylindrical with slightly inflated base,
up to 4-septate, sometimes faintly constricted at the septa, brown to fawn brown;
wall smooth and 1 μ thick; 33-81 [46] μ long, 3.5-5 μ wide at the base; terminating
in a phialide. Phialides lageniform, subhyaline to pale brown, 28-54 [38] μ long;
venter subcylindrical or rarely subellipsoidal, 12-29 [19] x 3.5-7 [5.3] μ; collarette
cylindrical, 12-32 [20] x 2-2.5 [2.3] μ; transition from venter to collarette abrupt;
ratio of mean lengths of collarette and venter = 1.05:1. Phialoconidia extruded
singly or in easily dispersible chains; cylindrical, ends blunt or slightly rounded,
unicellular, hyaline, 5.5-19 [11.5] x 1.5-2.5 [1.7] μ; mean conidium length/width
ratio = 6.7:1.

*Habitat: On Aesculus, Beilschmiedia tawa, B. taraire, Corylus avellana, Fagus sylva-
tica, Nothofagus solandri,* and *Quercus.*

12) *Chalara dictyoseptata.* Phialophores and conidia ex DAOM 93336a.

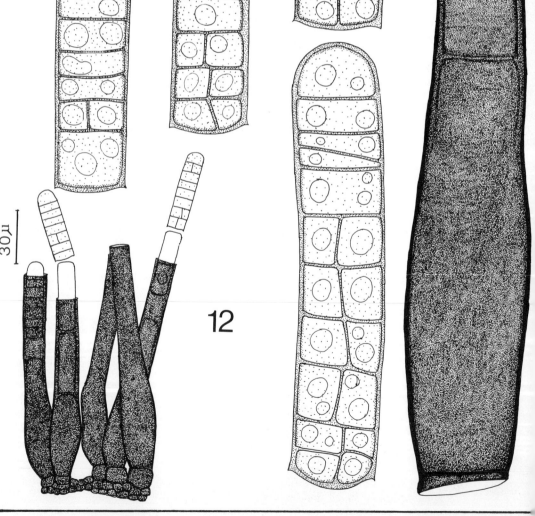

Chalara unicolor. Phialophore and conidia ex DAOM 110019.
Chalara inflatipes. Phialophore and conidia ex type in B.

B

13

14) *Chalara pteridina.* Phialophores and conidia ex IMI 27811.

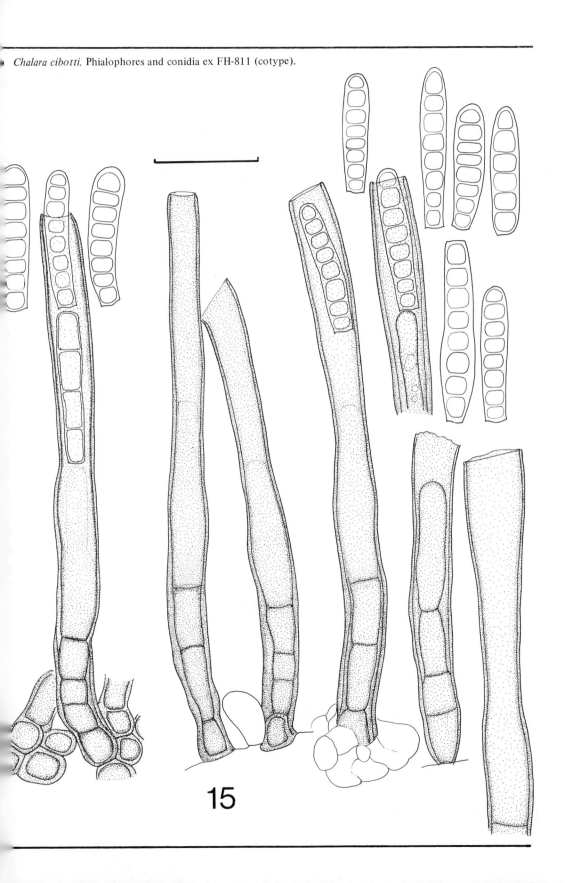

Chalara cibotti. Phialophores and conidia ex FH-811 (cotype).

15

16) *Chalara pulchra.* Phialophores and conidia ex DAOM 110020.

16

) *Chalara insignis.* Phialophores and conidia ex DAOM 28764.
) *Chalara bicolor.* Phialophores and conidia ex DAOM 110018.

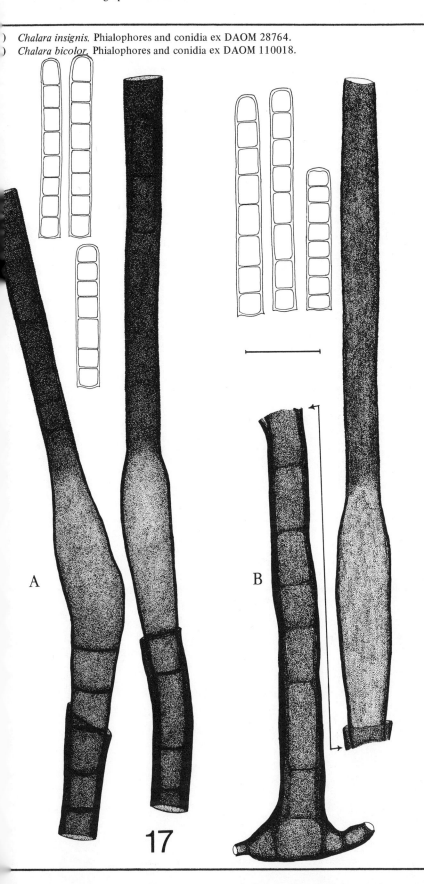

18A) *Chalara cladii.* Phialophores and conidia ex IMI 10171.
18B) *Chalara rubi.* Phialophores and conidia ex IMI 69882.

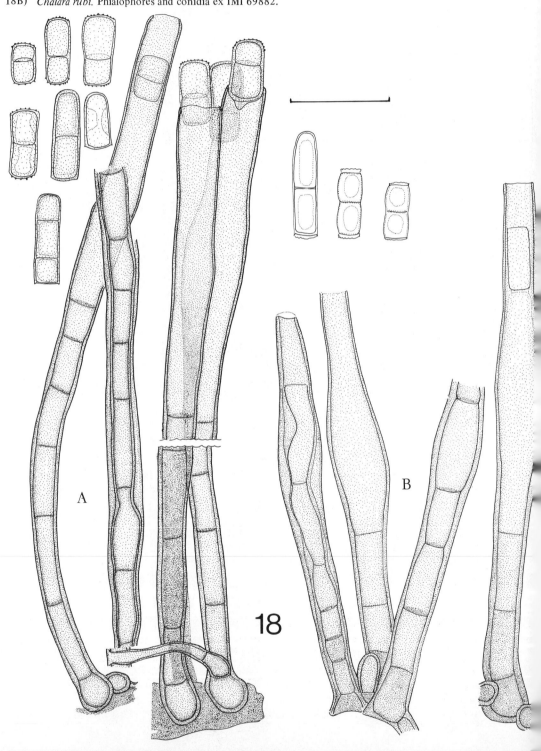

) *Chalara scabrida.* Phialophores and conidia ex PDD 32857.
) *Chalara emodensis.* Phialides and conidia ex IMI 93970b.
) *Chalara curvata.* Phialides and conidia ex PDD 32642.

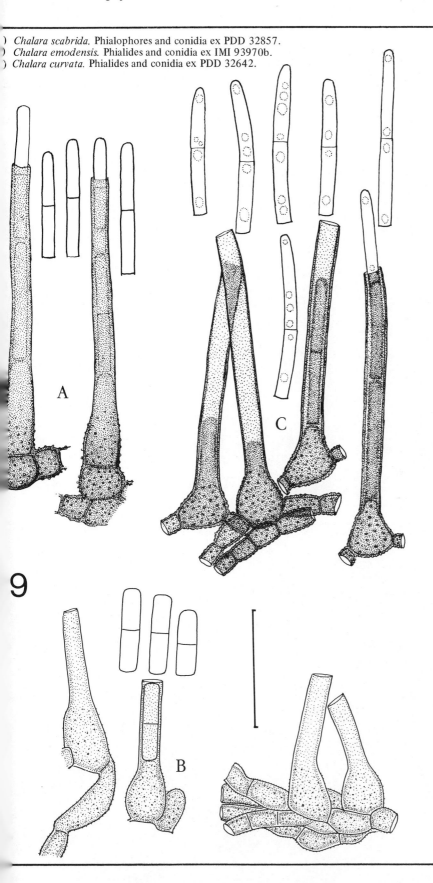

20A) *Chalara inaequalis.* Phialophores and conidia ex PDD 32643.
20B) *Chalara nigricollis.* Phialophores and conidia ex IMI 55278.
20C) *Chalara tubifera.* Phialophores and conidia ex PDD 32871.

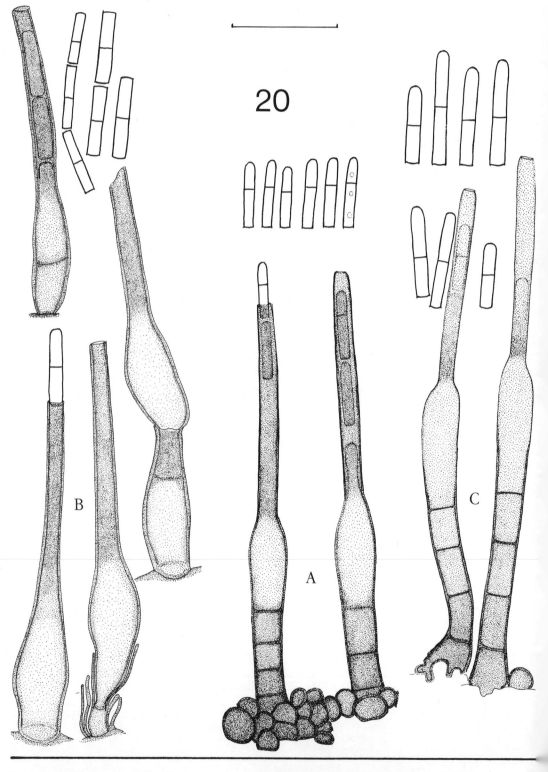

A) *Chalara agathidis.* Phialides and conidia ex PDD 32868.
B) *Chalara gracilis.* Phialophores and conidia ex PDD 32872.
C) *Chalara kendrickii.* Phialophores and conidia ex IMI 61712a.
D) Unnamed *Chalara* sp.
E & F) *Chalara angionacea.* Phialides and conidia ex PDD 32870 and 32869.

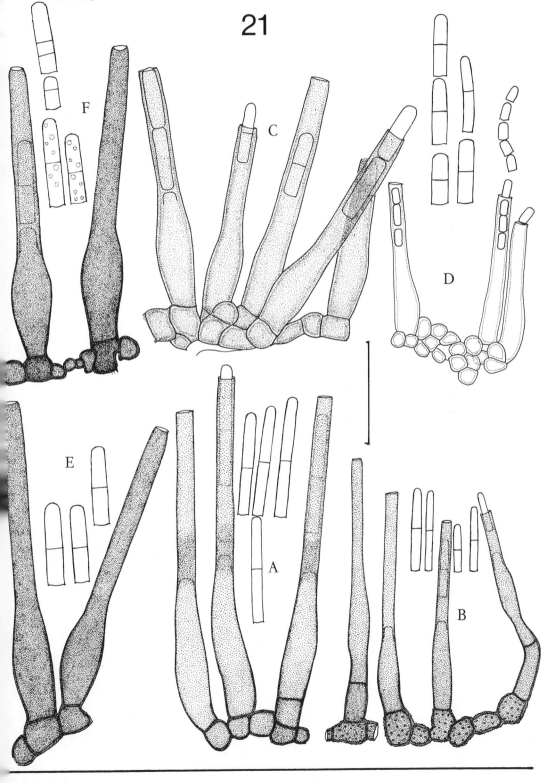

21

22A)   *Chalara acuaria.* Phialophores and conidia ex IMI 112288.
22B)   *Chalara rostrata.* Phialophores and conidia ex IMI 54897.

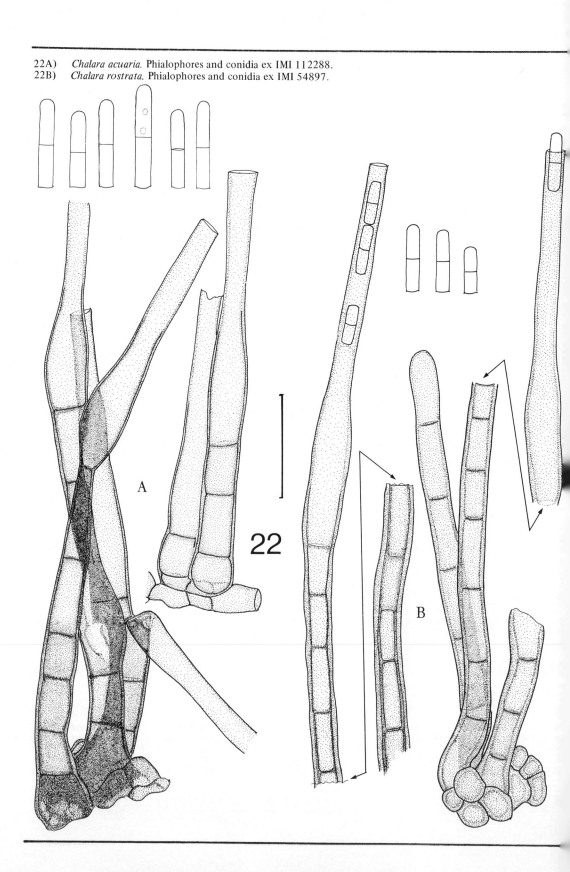

Specimens examined: 1) PAD 13/416 [*Holotype*] ex Herb. Saccardo, on rotting branches of *Quercus* sp.; 2) IMI 19219(k), on *Fagus sylvatica* cupules on ground, Swinton Pk., Yorks., U.K., 11.X.1947, S. J. Hughes; 3) IMI 19220, on *Corylus avellana* nut on ground, Swinton Pk., Yorks., U.K., 11.X.1947, S.J.H.; 4) IMI 19414 (c), on *Fagus sylvatica* cupules on ground, Ranmore Common, Surrey, U.K., 2.XI.1947, S.J.H.; 5) IMI 31452 (i), on *Aesculus* fruits, Swinton Pk., Yorks., U.K.,

A) *Chalara prolifera*. Phialophores and conidia ex IMI 111990.
B) *Chalara selaginellae*. Phialophores and conidia from type in BPI.

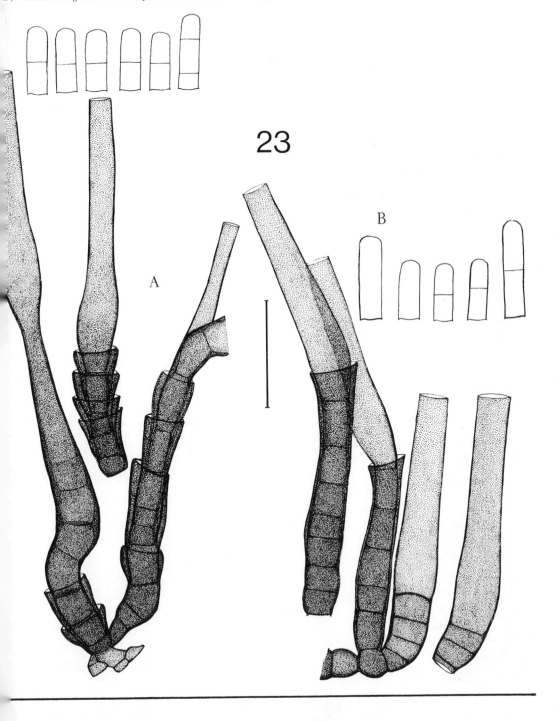

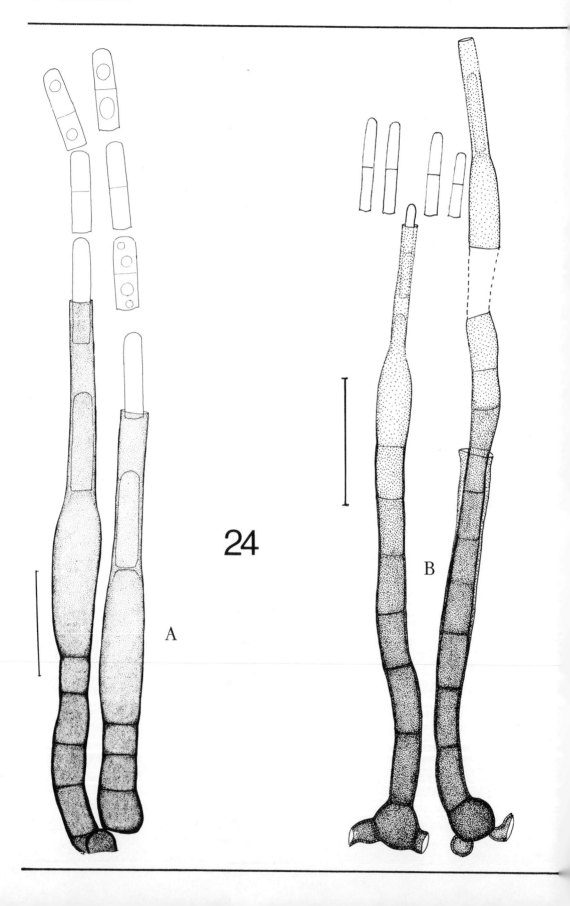

24

A

B

26.IX.1948, S.J.H.; 6) PDD 32875, on dead leaf of *Beilschmiedia tawa,* Orere Pt., Manukau Co., N.Z., 3.I.1974, B. Kendrick (KNZ 149); 7) PDD 32876, on dead leaf of *B. taraire,* Buck Taylor Tr., Lone Kauri Rd., Waitakere Ra., Auckland, N.Z., 22.XII.1974, B.K. (KNZ 35f); 8) PDD 32865, on dead leaf of *Nothofagus solandri,* L. Te Anau, Fiordland, N.Z., 13.IV.1974, B.K. (KNZ 613).

Known distribution: Italy, New Zealand, U.K.

C. affinis is close to *C. brevipes,* but has a usually subcylindrical rather than cylindrical or ellipsoidal venter, lacks a basally fuliginous collarette, and has longer and wider conidia that lack a basal marginal frill.

3) **Chalara agathidis** sp. nov. (Figure 21A)

Colonia effusa, pallide brunnea, pubescens. Mycelium vegetativum in stromata tenuia ex cellulis brunneis vel atrobrunneis, incrassatis, angularibus composita aggregatum. Phialophora 65-71 μ long., ad sessiles phialides redacta vel ex phialide terminali et brevi cellula fulcienti 6 x 6.5-7 μ composita. Phialides lageniformes, 61-67 (-71) [64] μ long., pallide brunneae, laeves; venter subcylindraceus, 22-28 [25] x 7-8 [7.5] μ; collum cylindraceum basaliter fuscum, 34-46 [39] x 3-3.5 [3.2] μ; transitio ex ventre ad collum gradatim; ratio long. colli et ventris =1.5:1. Phialoconidia in catenas breves extrusa; cylindracea, apice rotundato, base truncata, fimbriam marginalem ferente; 1-septata, hyalina, laevia; 17-24 [21] x 2.5-3 [2.7] μ; ratio conidii long./lat. = 7.7:1.

Colony effuse, pale brown, pubescent. Vegetative mycelium aggregated into thin stromata of brown to dark brown, thick-walled, angular cells. Phialophores 65-71 μ long, reduced to sessile phialides or composed of a terminal phialide and a short stalk cell, 6 x 6.5-7 μ. Phialides lageniform, 61-67 (-71) [64] μ long, pale brown, smooth; venter subcylindrical, 22-28 [25] x 7-8 [7.5] μ; collarette cylindrical, base dark; 34-46 [39] x 3-3.5 [3.2] μ; transition from venter to collarette gradual; ratio of mean lengths of collarette/venter = 1.5:1. Phialoconidia extruded in short chains; cylindrical, apex rounded, base truncate bearing a distinct marginal frill; 1-septate, hyaline, smooth; 17-24 [21] x 2.5-3 [2.7] μ; mean conidium length/width ratio = 7.7:1.

Habitat: On dead leaf of *Agathis australis.*

Specimen examined: PDD 32868 [*Holotype*], Ricker Tr., Piha Valley, Waitakere Ra., Auckland, N.Z., 9.II.1974, B. Kendrick (KNZ 380).

Known distribution: New Zealand.

Chalara agathidis superficially resembles *C. angionacea,* but its venter is longer and narrower, its collarette is narrower and darker at its base, and its conidia are longer and narrower.

4) **Chalara ampullula** (Sacc.) Sacc. (Figure 30D)
in Michelia, 1: 80, 1877.

$\equiv$ *Sporoschisma ampullula* Sacc.
in Atti Accad. scient. veneto-trent.-istriana, 4: 38,|1875.

= *Chalara aeruginosa* Höhn.
in Sber. Akad. Wiss. Wien, Abt. I, 111: 63, 1902.

Colony barely perceptible, superficial, effuse, whitish or cream yellow, downy to powdery. Phialophores reduced to phialides. Phialides borne directly on the

---

) *Chalara aotearoae.* Phialophore and conidia ex DAOM 110033b.
) *Chalara stipitata.* Phialophores and conidia ex PDD 32638.

25A) *Chalara rhynchophiala.* Phialides and conidia ex IMI 90620.
25B) *Chalara hughesii.* Phialophores and conidia ex DAOM 29354.

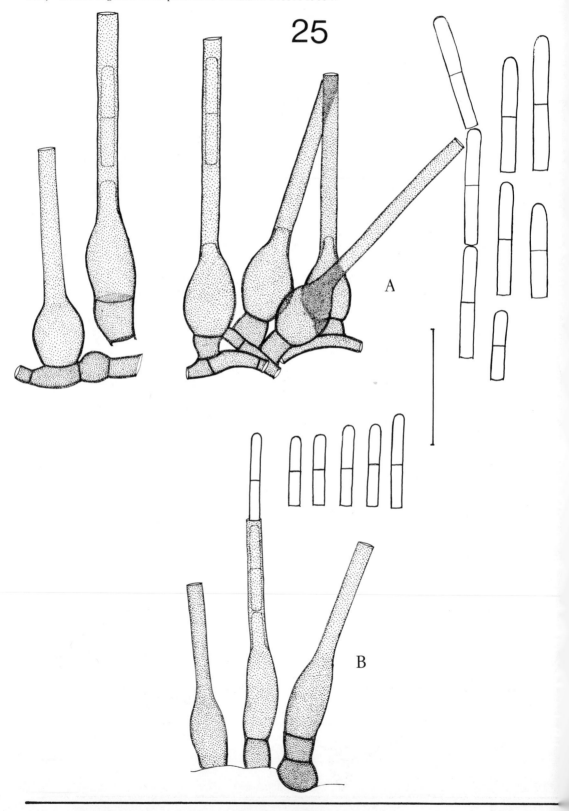

5A)   *Chalara urceolata.* Phialophores and conidia ex IMI 31315b.
5B)   *Chalara aurea.* Phialophores and conidia ex IMI 44550.

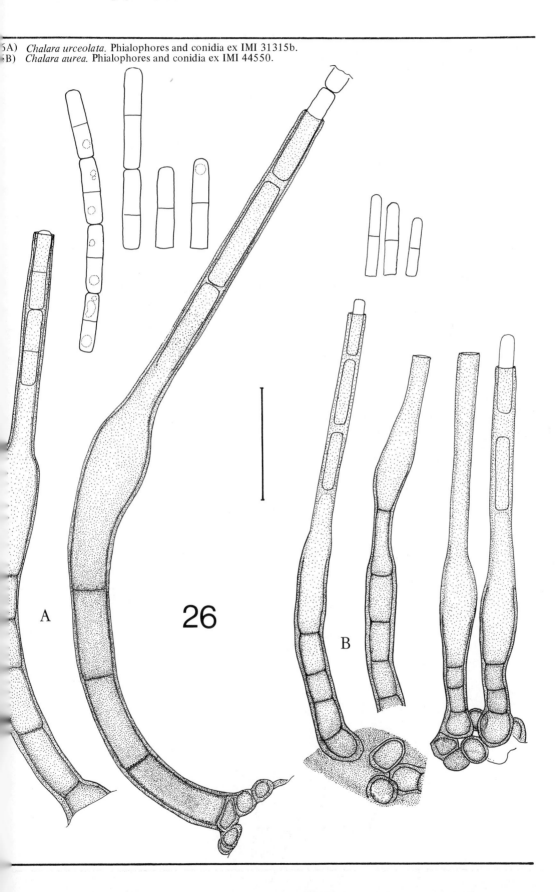

A   26   B

27) *Chalara spiralis.* Phialophores and conidia ex IMI 96708.

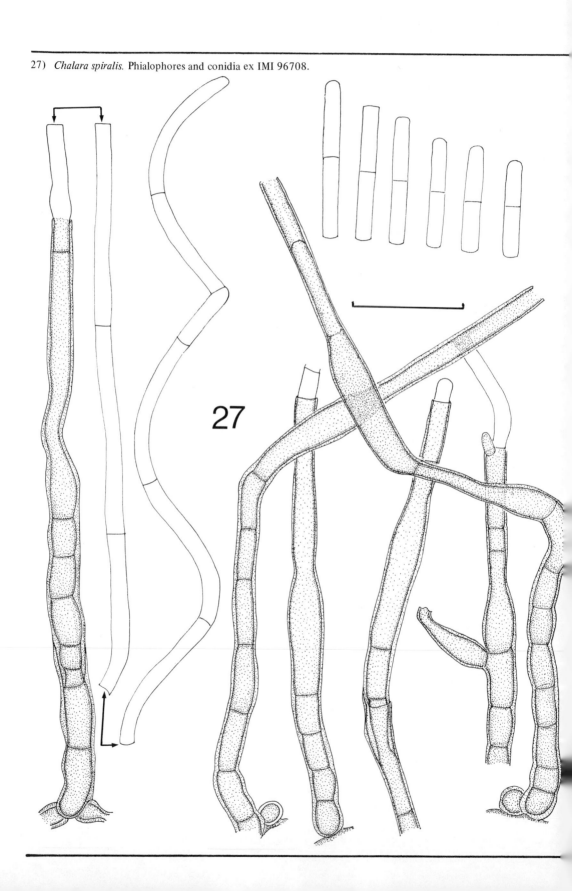

27

hyphae, sometimes scattered but usually clustered, conical or ampulliform, yellowish brown to pale brown, 11-20 [16] μ long; venter usually ellipsoidal, rarely subcylindrical or conical, 6.5-14 [9.9] x 4-6.5 [5.2] μ; collarette cylindrical, 3-10 [6] x 2-3 [2.5] μ; transition from venter to collarette abrupt, often marked by a pronounced constriction; ratio of mean lengths of collarette and venter = 0.6:1. Phialoconidia occurring singly or in easily dispersible (or sometimes persistent) chains; cylindrical with truncate or rounded ends, unicellular, hyaline, smooth-walled; 5-9 [7] x 1.5-2.5 [1.8] μ; mean conidium length/width ratio = 3.7:1.

Habitat: On *Gleditschia triacanthos*, *Salix* and *Vitis*.

Lectotype: Figure 33 of Saccardo in Michelia I, 1877.

Specimens examined: 1) FH 1614 in folder 11127, on *Salix*, Jaize, Bosnia, Yugoslavia, 1903, Höhnel; 2) FH 1624 in folder 11130, sub *C. sanguinea* 'in socio *C. aeruginosa*,' on rotting fruits of *Gleditschia triacanthos*, Osterr.; 3) FH 1611 in folder 11125, sub '*C. affinis*, Salix-zw-holz, Jaize, Bosnien, Osterr., 1903'; 4) B, 'in caudiis *Vitis viniferis*' [no other collection data].

Known distribution: Italy, Yugoslavia.

Typification: The type specimen of this species is not available in PAD, which is a repository for most of the fungi studied by Saccardo; nor is it available in any of the other major Italian herbaria consulted. There is little reason to expect that the specimen still exists. Saccardo illustrated the fungus in his Figure 33, published in Michelia 1, 1877, which offers useful diagnostic information. In accordance with Article 7, note 3 of the Code, this illustration is chosen as lectotype.

   C. ampullula is a remarkable marginal species that shows in the morphology of its phialides characters of *Phialophora* Medlar, from which it can be differentiated by its cylindrical conidia with blunt or truncate ends and formed in chains. It is close to *C. austriaca* and *C. fusidioides*. It can be distinguished from *C. austriaca* by its phialides, which are borne directly on the hyphae without any other supporting structure; and by the shape and size of its conidia. It differs from *C. fusidioides* in its relatively shorter, more or less obconical collarettes, its smaller phialides, and narrower conidia.

5) Chalara angionacea sp. nov. (Figure 21E, F)

   Colonia effusa, brunnea, pubescens. Mycelium vegetativum immersum, in stromata tenuia ex cellulis atrobrunneis, incrassatis composita aggregatum. Phialophora 60-65 μ long., ex atrobrunnea, usque ad 6.5 μ lat., fulcienti cellula, et phialide terminali, composita. Phialides lageniformes, 53-64 [57] μ long., pallide brunneae vel brunneae et concolorae, laeves; venter ellipsoideus, 16-23 [20] x 7.5-9 [8.2] μ; collum cylindraceum, 35-43 [39] x 3.5-4.5 [3.9] μ; transitione ex ventre ad collum gradatim; ratio long. colli et ventris = 1.9:1. Phialoconidia in catenas breves extrusa; cylindracea, apice rotundato; base truncata; 1-septata, hyalina, laevia; 13-18.5 [15] x 3-3.5 μ; ratio conidii long./lat. = 4.6:1.

   Colony effuse, brown, hairy. Vegetative mycelium immersed, aggregated into thin stromata of dark brown, thick-walled cells. Phialophores 60-65 μ long, made up of a dark brown stalk cell up to 6.5 μ wide, and a terminal phialide. Phialides lageniform, 53-64 [57] μ long, pale brown to brown and concolorous, smooth; venter ellipsoid, 16-23 [20] x 7.5-9 [8.2] μ; collarette cylindrical, 35-43 [39] x 3.5-4.5 [3.9] μ; transition from venter to collarette gradual; ratio of mean lengths of collarette/venter = 1.9:1. Phialoconidia extruded in short chains; cylindrical,

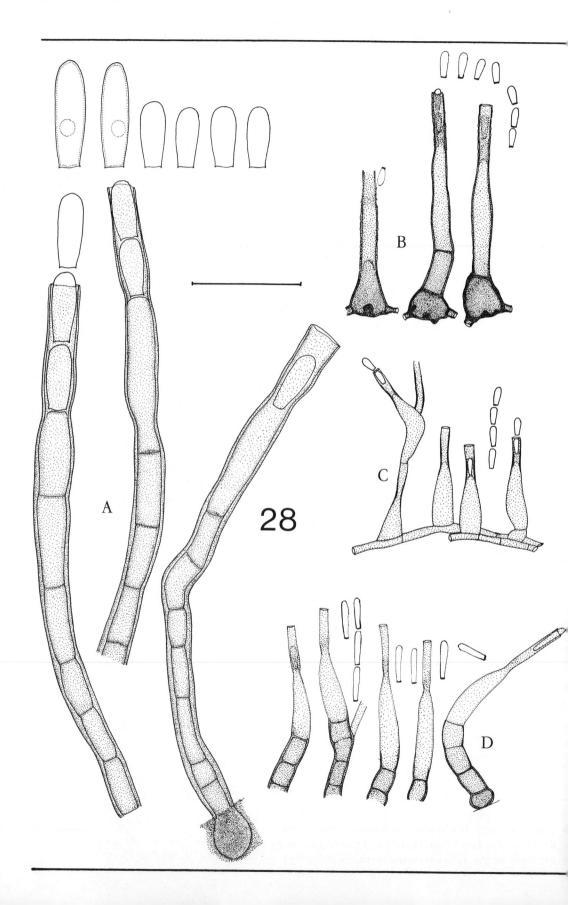

apex rounded, base truncate; 1-septate, hyaline, smooth; 13-18.5 [15] x 3-3.5 μ; mean conidium length/width ratio = 4.6:1.

Habitat: On dead leaves of *Beilschmiedia tawa* and *Knightia excelsa.*

Specimens examined: 1) PDD 32869, on dead leaf of *Knightia excelsa*, Orere Pt., Manukau Co., N.Z., 3.I.1974, B. Kendrick (KNZ 72); 2) PDD 32870 [*Holotype*], on dead leaf of *Beilschmiedia tawa*, Orere Pt., Manukau Co., N.Z., 3.I.1974, B.K. (KNZ 150).

Known distribution: New Zealand.
    Occasional aberrant conidia with up to 2 septa, and shorter 1-septate conidia, in each case with unequal cells, were seen in KNZ 72.
    Refer to *C. agathidis* for distinctions between it and *C. angionacea. C. angiona-cea* also resembles *C. kendrickii*, but differs from it by the less extensive basal aggregations of hyphal cells, its more bulbous venter, narrower collarette and longer conidia.

6) **Chalara aotearoae** Nag Raj & Hughes (Figure 24A)
    in N.Z. Jl. Bot., 12: 120, 1974.
    Colony effuse, black, hairy or densely tufted. Phialophores solitary to loosely aggregated, simple, erect, straight or variously bent, subcylindrical to cylindrical, 2-many-septate, dark brown at the base, becoming progressively paler above; wall smooth and not constricted at the septa; 65-280 [180] μ long, 6.5-8 μ wide at the base; terminating in a phialide. Phialides lageniform, pale brown, 59-69 (-72) [64] μ long; venter subcylindrical, 22-33 [28] x 8-10 (-12) [9.6] μ; collarette cylindrical, 31-42 [36] x 5-6.5 [5.8] μ; transition from venter to collarette abrupt; ratio of mean lengths of collarette and venter = 1.3:1. Phialoconidia extruded singly or in easily dispersible chains; cylindrical with rounded apex and truncate base bearing minute marginal frills, hyaline, 1-septate; wall smooth and unconstricted at the septum; 13-18 (-19) [16] x 3.5-5 [4.2] μ; mean conidium length/width ratio = 3.7:1.

Habitat: On *Aristotelia serrata.*

Specimen examined: DAOM 110033(b) ex *Holotype* in PDD 30405, Little Wanganui R., Westland, N.Z., 6.IV.1963, J. Dingley.

Known distribution: New Zealand.
    C. aotearoae resembles *C. cylindrosperma* and *C. stipitata*. It is distinct from *C. cylindrosperma* by its cylindrical or subcylindrical rather than ellipsoidal venter, and 1-septate conidia. It can be distinguished from *C. stipitata* by its larger phi-alides, longer and wider venters as well as collarettes, and wider conidia.

7) **Chalara aurea** (Cda.) Hughes (Figure 26B)
    in Can. J. Bot., 36: 747, 1958
    ≡ *Menispora aurea* Cda.
    in Icon. Fung., 2: 43, 1838.

) *Chalara breviclavata.* Phialophores and conidia ex DAOM 44962a.
) *Chalara brunnipes.* Phialophores and conidia ex PDD 32845.
) *Chalara* state of *Ceratocystis autographa.* Phialides and conidia ex type.
) *Chalara constricta.* Phialophores and conidia ex PDD 32854.

29A) *Chalara ellisii*. a. Funiculose mycelium bearing phialophores; b. Phialophores and conidia ex
        UAMH 1548.
29B) *Chalara cylindrica*. Phialophores and conidia ex IMI 69880.
29C) *Chalara bohemica*. Phialophores and conidia ex IMI 82786.

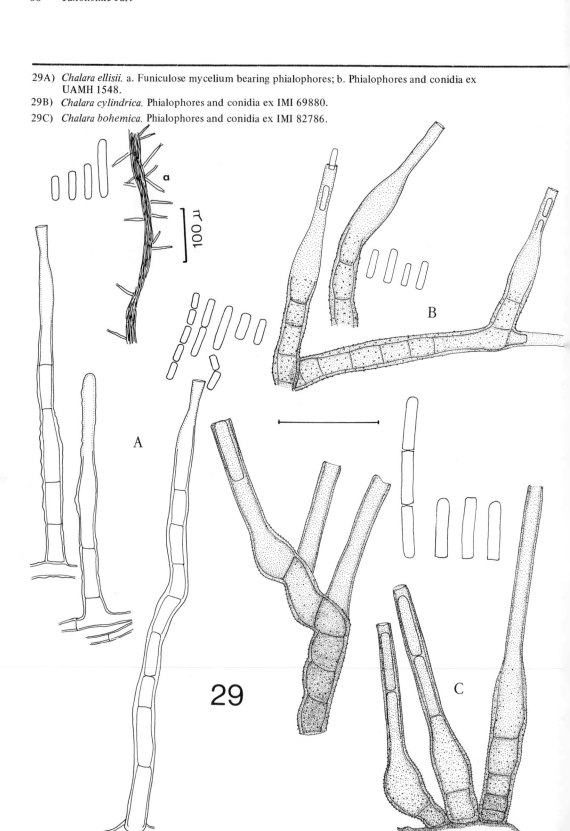

≡ *Cylindrocephalum aureum* (Cda.) Bon.
   in Handb. der Allgem. Mykologie, Stuttgart, p. 103, 1851.

   Colony superficial, effuse, yellowish white to golden yellow, velvety to hairy, forming irregular patches. Phialophores arising from a thin, stromatic layer of cells; simple, erect, cylindrical, up to 8-septate, with or without constrictions at septa, brown to yellowish brown, often lighter towards apex; walls smooth and 1 μ thick; 33-97 [59] μ long and 3.5-7.5 [5.1] μ wide at the base, terminating in a phialide. Phialides obclavate to lageniform, 29-76 [48] μ long; venter ellipsoidal, 8.5-27 [18] x 3.5-8.5 [6] μ; collarette cylindrical, 14-46 [29] x 2-4 [3] μ; transition from venter to collarette gradual; ratio of mean lengths of collarette and venter = 1.6:1. Proliferation uncommon, percurrent. Phialoconidia usually occurring in long, easily dispersible (or sometimes persistent) chains; cylindrical, often with rounded apex and truncate base, or sometimes both ends rounded; mostly 1-septate, rarely unicellular; hyaline, smooth-walled, 8-19 [13] x 2-2.5 [2.3] μ; mean conidium length/width ratio = 5.8:1.

Habitat: On *Aesculus, Betula, Carpinus, Ilex* and *Quercus.*

Specimens examined: 1) IMI 44550 (slide), ex *Holotype* in Herb. Corda. PR #515148; 2) IMI 19025(e), on *Quercus* wood chips on ground, Ranmore Common, Surrey, U.K., 25.V.1947, S. J. Hughes; 3) IMI 24898(b), inside bark of *Betula,* Richmond Pk., Surrey, U.K., 22.VIII.1948, S.J.H.; 4) IMI 60210, on dead branches of *Ilex europaeus,* Moulin Huet, Guernsey, U.K., 21.IX.1948, S.J.H.; 5) IMI 90619(a), on *Aesculus* mast, Kingthorpe, Pickering, Yorks., U.K., 1.XI.1961, W. G. Bramley; 6) FH 1611 in folder 11125, Herb. v. Höhn., Wiener Wald, Austria, 6.VII.1902, Höhnel; 7) FH 1611 in folder 11125, Herb. v. Höhn., Pfalzau, Wiener Wald, Austria, 21.V.1903, Höhnel; 8) FH 1611 in folder 11125, Herb. v. Höhn., Wiener Wald, Austria, 1.V.1904, Höhnel; 9) FH 1611 in folder 11125, on *Carpinus,* Wiener Wald, Austria, 24.IX.1904, Höhnel.

Known distribution: Austria, Czechoslovakia, U.K.
   Proliferation of phialides is uncommon and almost always percurrent through the tip of an old phialide or through the broken end of a conidiophore.
   C. aurea is like *C. spiralis* but differs from it in its wider, ellipsoidal rather than narrow, subcylindrical venter, and in its shorter and narrower conidia.

8) Chalara austriaca (Fautr. & Lamb.) comb. nov. (Figure 30B)
≡ *Chalara longipes* (Pr.) Cooke f. *austriaca* Fautr. & Lamb.
   in Revue mycol., 17: 69, 1895.

   Colony on the host not easily discernible. Phialophores often reduced to phialides, or cylindrical, up to 3-septate, subhyaline to pale brown; wall smooth and up to 0.5 μ thick; 12-27 μ long, 3-4 μ wide at the base. Phialides lageniform, subhyaline to pale brown, smooth-walled, 12-21 [16] μ long; venter subcylindrical, 6.5-13 [9.5] x 3.5-4 [3.7] μ; collarette cylindrical, 6.5-8 [6.8] x 2-2.5 [1.9] μ; transition from venter to collarette gradual; ratio of mean lengths of collarette and venter = 0.7:1. Phialoconidia extruded singly or in easily dispersible chains; cylindrical with blunt or truncate ends, unicellular, hyaline, smooth-walled; 4-6.5 [5] x 1-1.5 [1.4] μ; mean conidium length/width ratio = 3.5:1.

Habitat: On needles of *Pinus austriaca.*

Specimens examined: 1) UPS 1779 [*Holotype*], 1894, Fautrey; 2) 'C. Roumeguère Fungi Selecti exsiccati #6710, sur aiguilles de *Pinus austriaca,* été 1894, rec. cl. Dr. Lambotte,' in NY and FH.

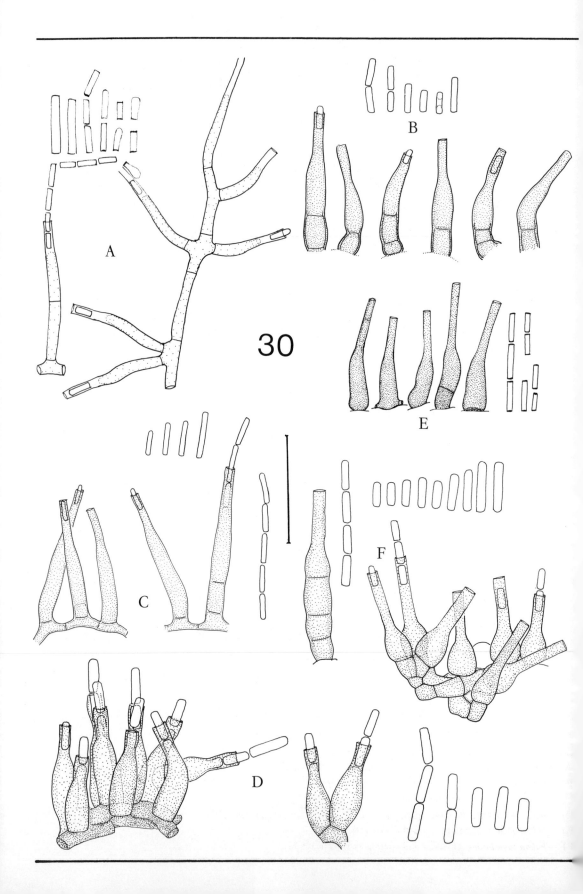

Known distribution: Ivory Coast, Africa.

There is very sparse material on the type specimen. The exsiccatum available in NY does not carry the fungus, while on the exsiccatum in FH very few intact conidiophores are discernible. These are up to 93 μ long, 4 μ wide at the base, up to 9-septate, brown at the base, pale brown or yellowish brown above, and smooth-walled. The phialides on these conidiophores are subcylindrical, 6 μ broad at the base, with a cylindrical collarette 18.5 x 2.5 μ. Only a few conidia were observed, and these were cylindrical with rounded ends, 1-septate, hyaline, smooth-walled, 12.5-17.5 x 2-2.5 μ. The median septum is barely perceptible. Roumeguère's disposition of the fungus under *C. longipes f. austriaca* was apparently an error. Since the exsiccati bear very little fungal material, they are of no taxonomic value and must be disregarded.

Because there are clearly perceptible differences in morphology between *C. longipes* and the taxon under consideration, the latter merits recognition as a distinct species.

Chalara austriaca resembles *C. constricta* and *C. microspora.* It differs from *C. microspora* in its smaller and narrower phialides, shorter collarettes and usually gradual rather than abrupt transition from venter to collarette and shorter conidia. It can be distinguished from *C. constricta* by its subcylindrical rather than ellipsoidal venter, collarette concolorous with venter, gradual rather than abrupt transition without a constriction in the wall, and narrower conidia.

9) Chalara bicolor Hughes (Figure 17B)
in N.Z. Jl. Bot., 12: 122, 1974.

Colony effuse, brown, hairy. Phialophores solitary and scattered, occasionally in groups of two to three, erect, straight or irregularly bent, cylindrical, with up to 12 septa, dark brown; wall smooth and not constricted at the septa; up to 170 μ long and 11-14 μ wide at the base, terminating in a phialide. Phialides subcylindrical to lageniform, 132-165 μ long; venter subcylindrical and slightly inflated, pale brown to brown, slightly to coarsely roughened, thin-walled, 50-55 x (12-) 14-16 (-18) μ, merging into a cylindrical, dark brown, thick-walled collarette, 7-9 μ wide at its base, broadening imperceptibly toward the apex, which is up to 11 μ wide: the distal part of the collarette is roughened, the roughnesses arranged in circumferential rows. Phialoconidia sometimes catenate; cylindrical, bluntly rounded at the apex and at the marginally frilled base, hyaline, 7-septate, smooth; (45-) 50-66 (-71) x 5.5-6 μ; mean conidium length/width ratio = 10:1.

Habitat: On decayed wood of *Podocarpus spicatus* and unidentified host.

Specimens examined: 1) On decayed wood of *Podocarpus spicatus,* Tongariro Nat. Pk., Erua, Wellington Pr., N.Z., 6.III.1963; 2) DAOM 110018 ex *Holotype* in PDD 30410, on unident. wood, Pureroa, Auckland Pr., N.Z. 21.III.1963, S. J. Hughes.

Known distribution: New Zealand.

C. bicolor is very close to *C. inflatipes* and *C. insignis* in possessing multiseptate, dark brown phialophores, and phialides with paler, rough-walled venters and dark brown collarettes with circumferential ridges along their length. *C. bicolor* may be

---

A) *Chalara* state of *Cryptendoxyla hypophloia* ex DAOM 147685.
B) *Chalara austriaca.* Phialophores and conidia ex type.
C) *Chalara microspora.* Phialides and conidia ex DAOM 51799.
D) *Chalara ampullula.* Phialides and conidia. ex FH 1624– sub *C. aeruginosa.*
E) *Chalara sessilis.* Phialides and conidia ex PDD 32639.
F) *Chalara fusidioides.* Phialides and conidia ex type in PR.

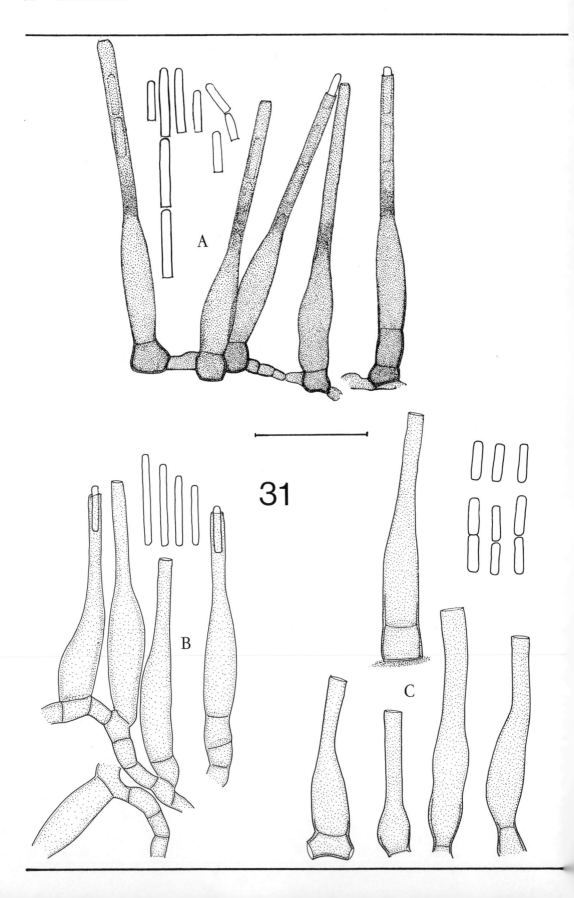

distinguished from *C. inflatipes* by its 7-septate and longer conidia, and from *C. insignis* by its longer, narrower conidia.

10) Chalara bohemica sp. nov. (Figure 29C)

Colonia superficialis, effusa, atrobrunnea, velutina. Phialophora simplicia, solitaria et dissita, vel in laxos fasciculos aggregata, cylindracea, recta vel varie flexa, 1-5-septata, subhyalina ad fumoso-brunnea, paries usque ad 1 μ cr., verrucosus et leniter constrictus ad septa, 21-72 [48] μ long.,4.5-8 [6.3] μ lat. ad basem, in phialidem terminantia. Phialides lageniformes, subhyalinae ad pallide brunneae, 15-59 [38] μ long.; venter ellipsoideus, pariete verrucoso, 6-18 [11] x 6-9.5 [7.5] μ; collum cylindraceum, pariete laevi, 15-39 [29] x 3-4.5 [3.5] μ; transitio ex ventre ad collum abrupta; ratio long. colli et ventris = 2.5:1. Phialoconidia singulatim vel in catenas breves extrusa; cylindracea, utrinque truncata vel obtusa, saepe tantum apice rotundato; unicellularia, hyalina, pariete laevi; 9.5-13 [11] x 2-3 [2.3] μ; ratio conidii long./lat. = 4.7:1.

Colony superficial, effuse, dark brown, velutinous. Phialophores simple, solitary and scattered or in lax clusters, cylindrical, erect or variously bent, 1-5-septate, subhyaline to smoky brown; wall up to 1 μ thick, verrucose and slightly constricted at the septa; 21-72 [48] μ long, 4.5-8 [6.3] μ wide at the base; terminating in a phialide. Phialides lageniform, subhyaline to pale brown, 15-59 [38] μ long; venter ellipsoidal, rough-walled, 6-18 [11] x 6-9.5 [7.5] μ; collarette cylindrical, smooth-walled, 15-39 [29] x 3-4.5 [3.5] μ; transition from venter to collarette abrupt; ratio of mean lengths of collarette and venter = 2.5:1. Proliferation common, sympodial, occasionally percurrent. Phialoconidia extruded singly or in short chains; cylindrical, both ends truncate or obtuse, often only the apex rounded; unicellular, hyaline, smooth-walled; 9.5-13 [11] x 2-3 [2.3] μ; mean conidium length/width ratio = 4.7:1.

Habitat: On stems of *Rubus fruticosus.*

Specimen examined: IMI 82786 [*Holotype*] , Skalnate Pleso (High Tatras), Czechoslovakia, 8.IX.1960, M. B. Ellis.

Known distribution: Czechoslovakia.

C. bohemica resembles *C. cylindrica* but has larger phialides, a rough-walled, wider venter which is shorter than the collarettes, longer and wider collarettes, larger conidia, and frequent sympodial proliferation of its phialides.

11) Chalara brachyspora Sacc. (Figure 34B)
in Michelia, 1: 81, 1877.

Phialophores solitary, scattered or moderately gregarious; simple, cylindrical to irregular, multi-septate, dark brown to reddish brown below, progressively paler above; wall smooth or occasionally verrucose in the basal part only; 41-220 [110] μ long, 4.5-7 [6] μ wide at the base, terminating in a phialide. Phialides lageniform or obclavate, pale brown, 23-42 [34] μ long; venter ellipsoidal, 12-18 [15] μ long and inflated at its submedian part to a width of 6.5-9.5 [7] μ; collarette cylindrical, 5-28 [19] x 3.5-5 [4.3] μ; transition from venter to collarette abrupt; ratio of mean lengths of collarette and venter = 1.25:1. Phialoconidia extruded singly or in persistent chains; short-cylindric, both ends rounded, or apex blunt and base truncate with a minute marginal frill; unicellular, hyaline, smooth-walled; 5-9.5 [7] x

) *Chalara brevipes.* Phialides and conidia ex PDD 32844.

) *Chalara affinis.* Phialophores and conidia ex IMI 19219k.

) *Chalara fungorum.* Phialides and conidia ex DAOM 43437.

32A) *Chalara quercina.* Phialophores and conidia ex type in BPI.
32B) *Chalara ungeri.* Phialophores and conidia ex IMI 20164.

32

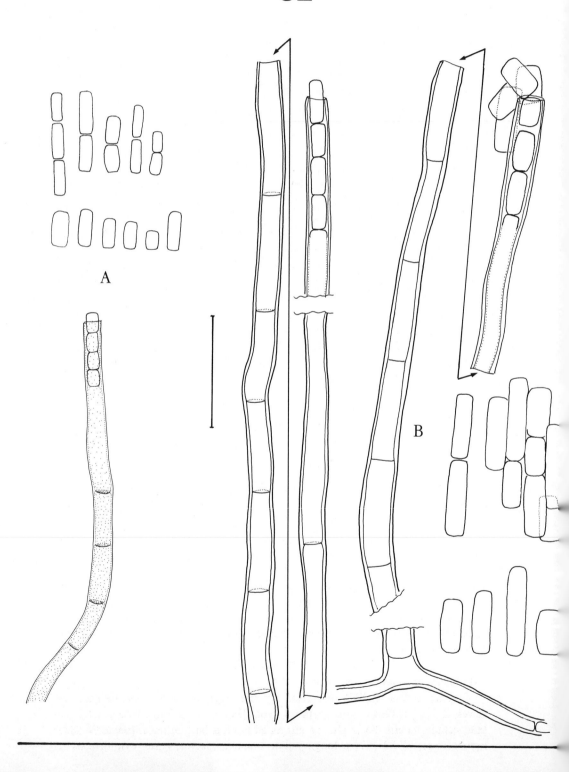

2.5-4 [3.2] μ; mean conidium length/width ratio = 2.2:1.

Habitat: On *Corylus* and unidentified wood.

Specimens examined: 1) DAOM 43438 (slide) ex *Holotype* in Herb. Saccardo in PAD, 'in Corylo silvo 73'; 2) BPI (sub. *C. brachyptera*), on wood, San Francisco, Calif., U.S.A., 26.III.1891, comm. Harkness.

Known distribution: Italy, U.S.A.

C. *brachyspora* resembles *C. cylindrosperma* and *C. germanica.* It can be distinguished from *C. cylindrosperma* by its often obclavate phialides, wider venters and collarettes, and shorter but wider conidia. If differs from *C. germanica* by virtue of its usually longer phialophores which do not arise from basal stromatic aggregations of vegetative hyphae, its often shorter venters, and its somewhat wider conidia.

12) Chalara breviclavata sp. nov. (Figure 28A)

Colonia superficialis, effusa, atrobrunnea, velutina. Phialophora simplicia, solitaria et sparsa vel in fasciculos laxos aggregata; cylindracea, recta vel varie flexa, usque ad 8-septata, atrobrunnea ad basem, supra pallescentia,pariete laevi usque ad 1 μ cr., sine constrictione ad septa, 71-115 [94] μ long., 4.5-6 [5.2] μ lat. ad basem, in phialidem terminantia. Phialides ampulliformes, subcylindraceae vel urceolatae, pallide brunneae vel paene subhyalinae ad apicem, 38-51 [46] μ long.; venter subcylindraceus, 20-33 [27] x 5.5-7 [6.3] μ; collum subcylindraceum vel obconicum, 15-20 [18] x 3.5-6.5 [4.8] μ; transitio ex ventre ad collum abrupta; ratio long. colli et ventris = 0.66:1. Phialoconidia singulatim vel interdum in catenas breves extrusa; ellipsoidea vel breviclavata, apice rotundato, base truncata fimbriam marginalem minutam sed perceptibilem ferente; unicellularia, hyalina, pariete laevi, 10-20 [14] x 3.5-4.5 [4.2] μ; ratio conidii long./lat. = 3.2:1.

Colony superficial, effuse, dark brown, hairy. Phialophores simple, solitary and scattered or in lax fascicles; cylindrical, erect, or variously bent, up to 8-septate, dark brown at the base, paler above, wall smooth, unconstricted at the septa, and up to 1 μ thick; 71-115 [94] μ long, 4.5-6 [5.2] μ wide at the base, terminating in a phialide. Phialides ampulliform, subcylindric or urceolate, pale brown to almost subhyaline at the apex, 38-51 [46] μ long; venter subcylindrical, 20-33 [27] x 5.5-7 [6.3] μ; collarette subcylindrical to obconical, 15-20 [18] x 3.5-6.5 [4.8] μ; transition from venter to collarette abrupt; ratio of mean lengths of collarette and venter = 0.66:1. Proliferation rare, percurrent. Phialoconidia occurring singly or sometimes in short chains; ellipsoidal to shortly clavate, rounded at the apex, truncate at the base with minute but discernible marginal frill, unicellular, hyaline, with smooth walls, 10-20 [14] x 3.5-4.5 [4.2] μ; mean conidium length/width ratio = 3.2:1.

Habitat: On wood.

Specimen examined: DAOM 44962(a) [*Holotype*] , on ? *Fraxinus,* nr. Old Chelsea, P.Q., Canada, 21.X.1954, S. J. Hughes.

C. *breviclavata* resembles *C. cibotti,* but can be distinguished by its narrower phialophores, shorter and narrower phialides, shorter collarettes and shorter, narrower, unicellular conidia.

13) Chalara brevipes sp. nov. (Figure 31A)

Colonia effusa, pallide brunnea. Hyphae vegetativae aggregatae, brunneae, laeves, 2.5 μ lat. Phialophora recta vel varie flexa, 49-60 μ long., basaliter 1-2-septata; cellula basalis 4-5 μ alt., et inflata ad 6-6.5 μ lat.; brunnea, laevia. Phialides lageniformes, 37-52 [47] μ long., pallide brunneae, laeves; venter cylindraceus vel

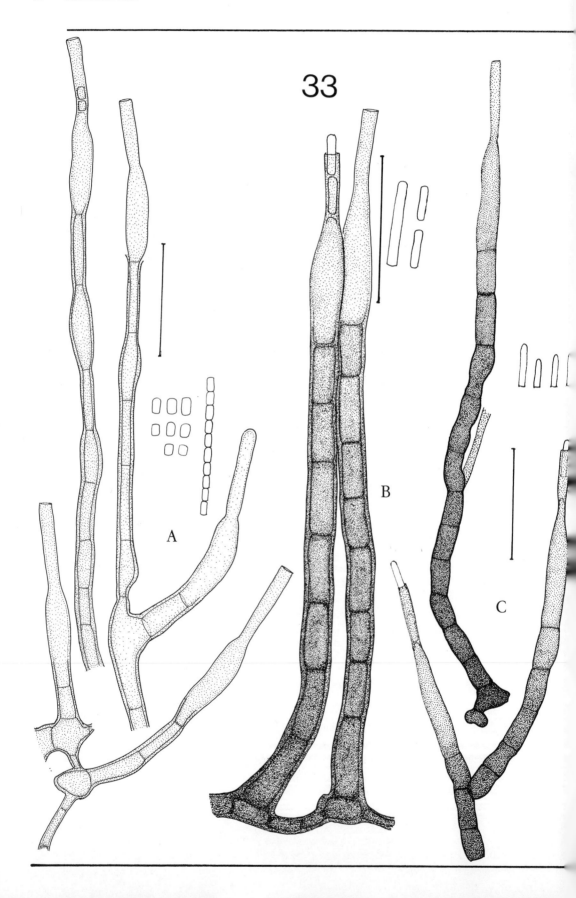

33

ellipsoideus; paries saepe parte mediana leniter recessus, 14-24 [20] x 5-7 [5.6] μ; collum cylindraceum, basaliter fuligineum, 23-31 [27] x 2-2.5 (-3) [2.3] μ; transitio ex ventre ad collum abrupta; ratio long. colli et ventris = 1.3:1. Phialoconidia in catenas breves extrusa, cylindracea; apice rotundato, base truncata, minutam fimbriam marginalem ferente; unicellularia, hyalina, laevia; 6-12 [8.9] x 1.5-2 μ; ratio conidii long./lat. = 5:1.

Colony effuse, pale brown. Vegetative hyphae aggregated, brown, smooth, 2.5 μ wide. Phialophores erect or variously bent; 49-60 μ long, basally 1-2-septate; basal cell 4-5 μ high and swollen to 6-6.5 μ wide; brown, smooth. Phialides lageniform, 37-52 [47] μ long, pale brown, smooth; venter cylindrical or ellipsoidal, wall often slightly invaginated in the median part; 14-24 [20] x 5-7 [5.6] μ; collarette cylindrical, basally fuliginous; 23-31 [27] x 2-2.5 (-3) [2.3] μ; transition from venter to collarette abrupt; ratio of mean lengths of collarette/venter = 1.3:1. Phialoconidia extruded in short chains; cylindrical, apex rounded, base truncate with a minute marginal frill; unicellular, hyaline, smooth, 6-12 [8.9] x 1.5-2 μ; mean conidium length/width ratio = 5:1.

Habitat: On dead leaves of *Podocarpus dacrydioides.*

Specimen examined: PDD 32844 [*Holotype*] , Big Tree Tr., Peel Forest Pk., Canterbury Prov., N.Z., 21.IV.1974, B. Kendrick (KNZ 677).

For affinities refer to *C. affinis.*

14) Chalara brevispora sp. nov. (Figure 33A)

Phialophora simplicia, recta vel varie flexa, cylindracea, multiseptata, septa superne obscura, atro-brunnea vel fumoso-brunnea, pariete laevi et 1 μ cr., sine constrictione ad septa, 41-145 [81] μ long., 3-5 [3.7] μ lat. ad basem, in phialidem terminantia. Phialides obclavatae vel lageniformes, pallide brunneae, 19-45 [35] μ long.; venter aliquantum cylindraceus, 10-24 [19] x 3.5-6 [4.5] μ; collum cylindraceum, 9-20 [16] x 2-3 [2.6] μ; transitio ex ventre ad collum abrupta; ratio long. colli et ventris = 0.9:l. Phialoconidia vulgo in catenas, interdum singulatim, extrusa; cylindracea, utrinque rotundata, vel obtusa, unicellularia, hyalina, pariete laevi, 2-4 [2.8] x 1.5-2 [1.8] μ; ratio conidii long./lat. = 1.5:1.

Phialophores simple, erect, straight or variously bent, cylindrical, many-septate, septa in the upper part indistinct; dark brown or smoky brown; wall smooth, 1 μ thick, and not constricted at septa; 41-145 [81] μ long, 3-5 [3.7] μ wide at the base, terminating in a phialide. Phialides obclavate to lageniform, light brown, 19-45 [35] μ long; venter cylindrical, 10-24 [19] x 3.5-6 [4.5] μ; collarette cylindrical, 9-20 [16] x 2-3 [2.6] μ; transition from venter to collarette abrupt; ratio of mean lengths of collarette and venter = 0.9:1. Phialoconidia sometimes occurring singly, but mostly in chains; cylindrical or short-cylindrical, ends rounded or blunt, unicellular, hyaline, smooth-walled, 2-4 [2.8] x 1.5-2 [1.8] μ; mean conidium length/width ratio = 1.5.1.

Habitat: On *Quercus* sp.

Specimen examined: DAOM 71539(b) (slide) [*Holotype*] , Ottawa, Ont., Canada 30.X.1960, G. L. Hennebert 1583-2.

) *Chalara brevispora.* Phialophores and conidia ex DAOM 71359b.
) *Chalara longipes.* Phialophores and conidia ex type in B.
) *Chalara novae-zelandiae.* Phialophores and conidia ex PDD 32848.

34A) *Chalara germanica.* Phialophore and conidia ex type in B.
34B) *Chalara brachyspora.* Phialophores and conidia ex DAOM 43438.
34C) *Chalara parvispora.* Phialophores and conidia ex DAOM 110036.

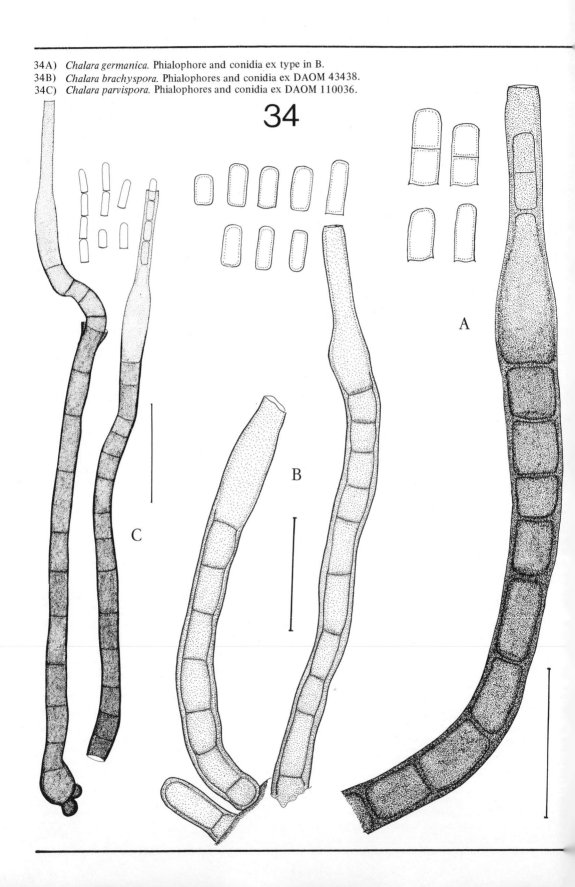

Dried agar cultures of the fungus have been examined. Colonies on PDA (27.1.61-21.3.61) 21 mm diam., velvety, grayish brown, with deeply indented margins. On MA (27.1.61-21.3.61) colonies 70 mm diam., velvety to lax cottony at the periphery, grayish brown with dendroid margins. Phialophores as observed in nature, but 32-165 [88] μ long, 2-5 [3.2] μ wide at the base; phialides 19-45 [28] μ long, venter 3.5-5 [4.5] μ wide, and 2-3.5 μ wide at the apex of the collarette. Phialoconidia 2.5-5 [3.2] x 1.5-2 [1.7] μ, accumulating in large slimy heads.

Meyer (1959) described an unnamed species of *Chalara* isolated from soil in the Republic of Zaire. His description went as follows: "Colonie gris noirâtre, laineuse, reverse foncé ou noir; mycelium immergé hyalin à subhyalin. Phialides directement insérées sur les hyphes fertiles, foncées, de 40 à 70 μ présentant un léger renflement vers le quart supérieur, de 3 à 4 μ de diamètre à la base, de 4.5 à 5 μ au niveau du renflement et de 3 μ vers l'extrémitaté. Phialospores hyalines à subhyalines, en chaîne basipète, cylindriques, délimitées dans la phialide; on y trouve souvent deux ou trois conidies préformées avec un corps réfringent central, 1.8 à 2.2 x 3.5 à 4 μ. Les chaînes de spores sont droites ou enroulées sur elles-mêmes."

We have not been able to study Meyer's material, but his description and illustration of the fungus match *C. brevispora*.

The short cylindrical conidia borne in usually obclavate phialides, terminating the often nodulose phialophores, set *C. brevispora* apart from other species of this group.

15) Chalara brunnipes sp. nov. (Figure 28B)

Colonia effusa, dissita, velutina. Stroma nullum. Phialophora sparsa vel gregaria, curta, recta vel flexa, 39-45 μ long., 0-2-septata; cellula basali subglobosa vel late conica, brunnea vel fuliginea, interdum aspera, 4.5-8 μ long. et 6-9 μ lat., praedita. Phialis terminalis subcylindracea, 25-34 [30] μ long., pallide brunnea, saepe ad colli basem leniter fuscescens; venter subcylindraceus, 14-19 [16.5] x 3.5-5.5 [4.2] μ; collum plerumque anguste obconicum, interdum cylindraceum, minute asperum, 11-15 [13.6] x 1.5-2.5 [2] μ; transitio ex ventre ad collum gradatim; ratio long. colli et ventris = 0.8:1. Phialoconidia in catenas longas extrusa; breviclavata, base truncata; unicellularia, hyalina, laevia, 3.5-5 [4.3] x 1.5 μ; ratio conidii long./lat. = 2.8:1.

Colony effuse, scattered, hairy. Stroma lacking. Phialophores scattered to gregarious, short, erect or bent, 39-45 μ long, 0-2-septate, with a subglobose to broadly conical, brown or smoky brown, sometimes asperate, basal cell, 4.5-8 μ long and 6-9 μ wide. Terminal phialide subcylindrical, 25-34 [30] μ long, pale brown, often slightly darker at the base of the collarette; venter subcylindrical, 14-19 [16.5] x 3.5-5.5 [4.2] μ; collarette usually narrow-obconic, occasionally cylindrical, minutely asperate, 11-15 [13.6] x 1.5-2.5 [2] μ; transition from venter to collarette gradual; ratio of mean lengths of collarette and venter = 0.8:1. Phialoconidia extruded in long chains; short clavate with a truncate base, unicellular, hyaline, smooth, 3.5-5 [4.3] x 1.5 μ; mean conidium length/width ratio = 2.8:1.

Habitat: On dead leaves of *Nothofagus fusca*.

Specimens examined: 1) PDD 32845 [*Holotype*], Cascade Falls Tr., Orewera Nat. Pk., N.Z., 24.III.1974, B. Kendrick (KNZ 526); 2) PDD 32846, L. Waikereite Tr., Orewera Nat. Pk., N.Z., 23.III.1974, B. Kendrick (KNZ 548).

Known distribution: New Zealand.

The broadly conical fuliginous basal cell is characteristic of *C. brunnipes*, which is similar to *C. austriaca* and *C. constricta*. It is distinct from *C. austriaca* in its

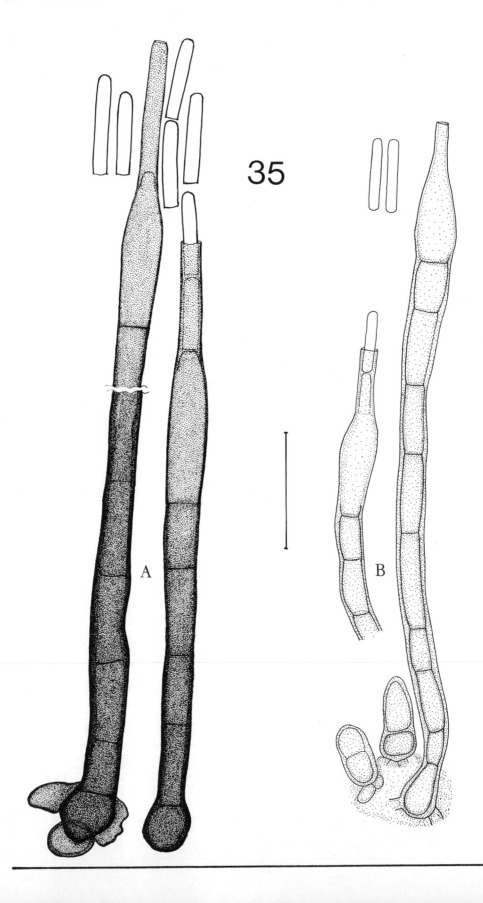

35

longer phialides, longer venter, minutely asperate collarettes and short clavate, rather than cylindrical, conidia. It can be distinguished from *C. constricta* by its subcylindrical, rather than lageniform, and longer phialides; subcylindrical and longer venter, minutely asperate collarettes, gradual rather than abrupt transition, and shorter, narrower conidia.

16) Chalara cibotti (Plunk.) comb. nov. (Figure 15)
≡ *Excioconidium cibotti* Plunk. apud Stevens
  in Bull. Bernice P. Bishop Mus., 19: 156, 1925.

Colony superficial, effuse, brown, hairy. Phialophores simple, solitary or in small clusters, erect, cylindrical, dark brown to reddish brown, 3-4-septate, smooth, not constricted at the septa, 83-125 μ long, 5.5-9 [7] μ wide at the base, terminating in a phialide. Phialides obclavate or urceolate, 56-86 [70] μ long; venter cylindrical or subcylindrical, 15-28 [23] x 6-12 [9] μ; collarette cylindrical or obconical, 32-57 [47] x 4.5-8 (-12) [7] μ; transition from venter to collarette gradual or abrupt; ratio of mean lengths of collarette and venter = 2:1. Phialoconidia rather cuneiform or obovoid with a broad, rounded apex, and a narrow, truncate base with marginal frill; up to 7-septate, hyaline, with smooth, thick walls, not constricted at the septa; 24-36 [28] x 4.5-6.5 [5.4] μ; mean conidium length/width ratio = 5:1.

Habitat: On *Cibotium chamissoi.*

Specimens examined: 1) FH 811 [*Syntype*], on *Cibotium chamissoi*, Kilauea, Hawaii, U.S.A., 13.VII.1921, F. L. Stevens; 2) DAOM 114330 (ex Herb. MSC), on dead rachis of *Cibotium chamissoi*, Waikeau Trail, Hawaii, U.S.A., 23.III.1940, E. A. Bessey #557.

Known distribution: Hawaii.
For affinities see *C. breviclavata.*

17) Chalara cladii M. B. Ellis (Figure 18A)
in Mycol. Pap., 79: 21, 1961.

Colony effuse, reddish brown or dark brown, velvety. Phialophores solitary or aggregated; simple, cylindrical, multi-septate, brown to dark brown; wall smooth and 1 μ thick; not constricted at septa, 110-610 [270] μ long, 3.5-7.5 [5.3] μ wide at the base, terminating in a phialide. Phialides subcylindrical, pale brown, 28-62 [50] μ long, narrow at the base, 6.5-11 [9] μ wide in the median part and 4.5-11 [7] μ wide at the apex; transition from venter to collarette almost imperceptible. Phialoconidia occurring singly; subcylindrical, broad and truncate at the base, slightly narrowed and rounded or truncate at the apex, mostly 1-septate but occasionally unicellular or 2-septate, cells unequal, wall sometimes slightly constricted at the septa, end walls verrucose, side walls smooth or slightly verrucose at each end; hyaline to pale brown, 8-18 [14] x 5-7 [6] μ; mean conidium length/width ratio = 2.3:1.

Habitat: On dead leaves of *Cladium mariscus.*

Specimens examined: All on *Cladium mariscus:* 1) IMI 10171 [*Holotype*], Wheatfen Broad, Norfolk, U.K., I.1947, E. A., M. B. and J. P. Ellis; 2) IMI 21399a,

A) *Chalara nothofagi.* Phialophores and conidia ex PDD 32847.
B) *Chalara cylindrosperma.* Phialophores and conidia ex IMI 44549.

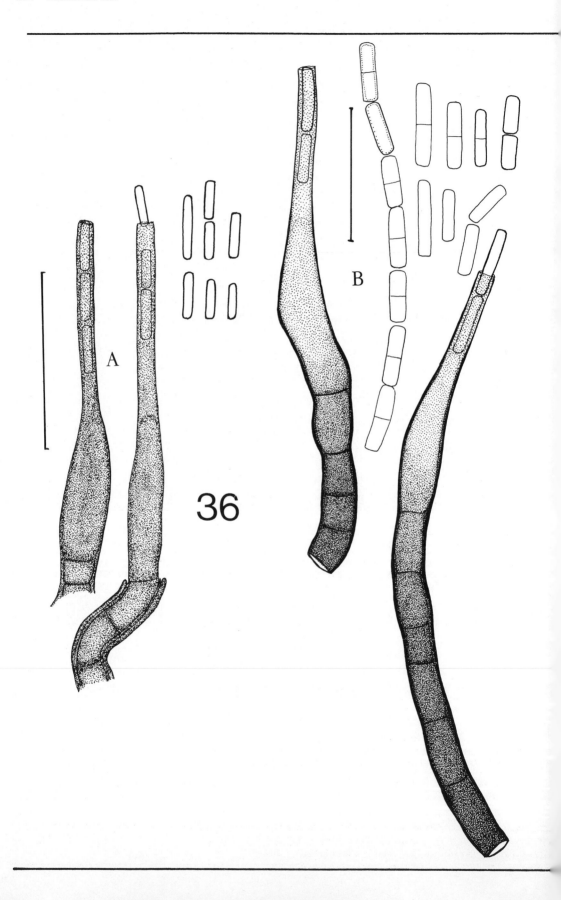

A

36

B

III.1940; 3) IMI 14875f; 4) IMI 16534d, IV.1947; 5) IMI 15393b, V.1947; 6) IMI 17355b, VIII.1947; 7) IMI 27789, 1948; 8) IMI 32602, XII.1948; 9) IMI 34918a, 1949.

Known distribution: U.K.

C. cladii is unique among *Chalara* species in possessing subcylindrical, usually septate, pigmented conidia with partly verrucose walls.

18) Chalara constricta sp. nov. (Figure 28D)

Colonia inconspicua, pallide brunnea. Phialophora solitaria vel aggregata, recta vel varie flexa, usque ad 42 μ long., et 4 μ lat. ad basem, ex stipite cylindrico, pallide brunneo, laevi, obscure 1-4-septato, et phialide terminali composita. Phialides lageniformes, 17-26 [22] μ long.; venter ellipsoideus, 9-16 [12.5] x 3-4 [3.7] μ; collum cylindraceum, basaliter fuligineum, 7-11 [8.3] x 1.5-2 μ; transitio ex ventre ad collum abrupta et parietis constrictione notata; ratio long. colli et ventris = 0.65:1. Phialoconidia catenulata, anguste clavata, base truncata, apice rotundato et leniter dilatato, unicellularia, hyalina, laevia, 3.5-8 [6.2] x 1-2 [1.5] μ; ratio conidii long./lat. = 4:1.

Colony inconspicuous, pale brown. Phialophores solitary or aggregated, straight or variously bent, up to 42 μ long and 4 μ wide at the base, composed of a cylindrical stalk which is pale brown, smooth-walled, indistinctly 1-4-septate; with a terminal phialide. Phialides lageniform, 17-26 [22] μ long; venter ellipsoidal, 9-16 [12.5] x 3-4 [3.7] μ; collarette cylindrical, basally fuliginous, 7-11 [8.3] x 1.5-2 μ; transition from venter to collarette abrupt and marked by a constriction in the wall; ratio of mean lengths of collarette and venter = 0.65:1. Phialoconidia catenulate, narrowly clavate to cylindrical; base truncate, apex rounded, slightly wider; unicellular, hyaline, smooth-walled; 3.5-8 [6.2] x 1-2 [1.5] μ; mean conidium length/width ratio = 4:1.

Habitat: On leaves of *Agathis australis.*

Specimens examined: 1) PDD 32873, Kauri Glen Pk., Northcote, Auckland, N.Z., 19.II.1974, B. Kendrick (KNZ 360); 2) PDD 32856, Unmarked Tr. off Scenic Dr., Waitakere Ra., Auckland, N.Z., 27.II.1974, B.K. (KNZ 430); 3 & 4) PDD 32855, 32851, Summit Tr., Little Barrier Is., N.Z., 7.III.1974, B.K. (KNZ 482b, 495); 5) PDD 32854 [*Holotype*], nr. stream S. of Te Wairere, Little Barrier Is., N.Z. 9.III.1974, B.K. (KNZ 456); 6 & 7) PDD 32640, 32853, Atkinson's Bush, Titirangi, Auckland, N.Z., 15.V.1974, B.K. (KNZ 624, 626); 8) PDD 32850, Large Kauri Tr., Waitakere Ra., Auckland, N.Z., 14.II.1974, B.K. (KNZ 327); 9) PDD 32852, Kauri Pk., Scenic Dr., Waitakere Ra., Auckland, N.Z., 15.V.1974, B.K. (KNZ 729b).

Known distribution: New Zealand.

For affinities refer to *C. austriaca.*

19) Chalara crassipes (Pr.) Sacc. (Figure 36A)

in Sylloge Fung., 4: 335, 1886; Nag Raj & Kendrick
in Can. J. Bot., 49:2119, 1971.
≡ *Cylindrosporium crassipes* Pr.
  in Linnaea, 24: 106, 1851.

) *Chalara crassipes.* Phialophores and conidia ex type in B.
) *Chalara ginkgonis.* Phialophores and conidia ex type.

37) *Chalara* state of *Ceratocystis adiposa.* A. conidia and chlamydospores ex IMI 121285. B. Phialophores, conidia and chlamydospores ex IMI 99465 from culture on *Pinus strobus* board. C. Chlamydospores ex IMI 21355.

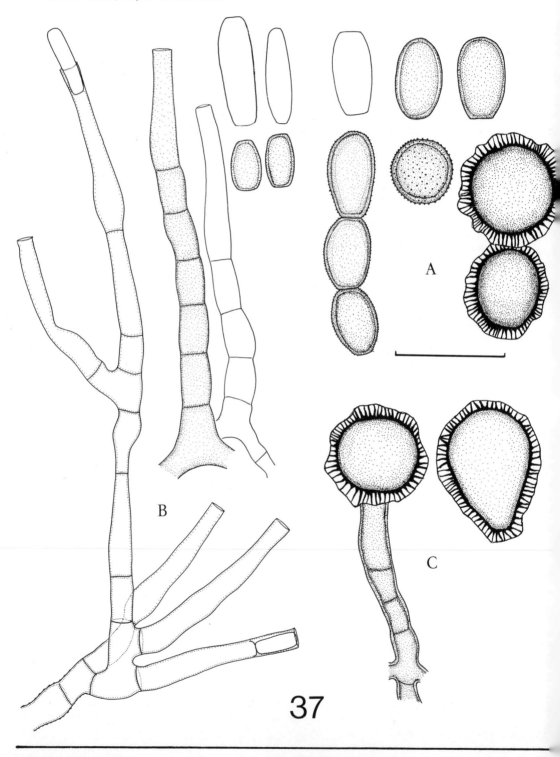

≡ *Chalara crassipes* (Pr.) Lindau
   in Rabenh. Kryptogamenflora, 8: 754, 1907.

Colony superficial, effuse, brown, velutinous. Phialophores simple, crowded, cylindrical, sparsely septate, brown to dark brown below, pale brown near the phialide, wall smooth and not constricted at the septa, 27-67 [45] μ long, 3-4.5 [3.7] μ wide at the base, terminating in a phialide. Phialides lageniform, obclavate, subhyaline to pale brown, 18-42 [32] μ long; venter cylindrical, occasionally ellipsoidal, 8-27 [15] x 4.5-5.5 [4.8] μ; collarette cylindrical, 9-24 [19] x 1.5-2.5 [1.7] μ; transition from venter to collarette gradual; ratio of mean lengths of collarette and venter = 1.25:1. Phialoconidia cylindrical with obtuse ends, unicellular, hyaline, smooth-walled; 4.5-6.5 [5.2] x 1-1.5 [1.3] μ; mean conidium length/width ratio = 4:1; conidial chains not observed.

Habitat: On conifer wood.

Specimen examined: B 229 (mappe 230c) [*Lectotype*] no other collection data.

Known distribution: Germany.
   Chalara crassipes resembles *Chalara affinis* but has smaller, narrower phialides, and correspondingly shorter, narrower conidia.

20) Chalara curvata sp. nov. (Figure 19C)

Colonia effusa, atra, caespitosa vel lanata. Mycelium superficiale aggregatum; hyphae 3.5-4 μ lat., pallide brunneae vel brunneae, parietibus asperis. Phialophora ad phialides simplices et sessiles redacta et ex cellulis mutatis hypharum superficialium enascentia. Phialides lageniformes, 35-50 [44] μ long., pallide brunneae et concolorae; venter brevis et asper, late conicus vel subglobosus; 6-8 [7.1] x 6-9 [7.4] μ; collum multo longius, cylindraceum, laeve, et saepe leniter curvata; 28-41 [38] x 3-3.5 μ; transitio ex ventre ad collum abrupta; ratio long. colli et ventris = 5.3:1. Phialoconidia singulatim extrusa, cylindracea, apice rotundato et base truncata minutam fimbriam marginalem ferente, recta vel leniter curvata, 1-septata, hyalina, laevia, 22-30 [26.5] x 2.5-3 (-3.5) [2.7] μ; ratio conidii long./lat. = 10:1.

Colony effuse, caespitose to woolly, black. Superficial mycelium aggregated, hyphae 3.5-4 μ wide, pale brown to brown with asperate walls. Phialophores reduced to simple, sessile phialides arising from modified cells of the superficial hyphae. Phialides lageniform, 35-50 [44] μ long, pale brown and concolorous throughout; venter short, asperate, broadly conical or subglobose; 6-8 [7.1] x 6-9 [7.4] μ; collarette much longer, cylindrical, often slightly curved, smooth-walled; 28-41 [38] x 3-3.5 μ; transition from venter to collarette abrupt; ratio of mean lengths of collarette/venter = 5.3:1. Conidia extruded singly; cylindrical, apex rounded, base truncate with a minute marginal frill; straight or slightly curved, 1-septate, hyaline, smooth-walled, 22-30 [26.5] x 2.5-3 (-3.5) [2.7] μ; mean conidium length/width ratio = 10:1.

Habitat: On leaf of *Dracophyllum traversii*

Specimen examined: PDD 32642 [*Holotype*], Scott's Tr., Arthur's Pass Nat. Pk., N.Z., 27.IV.1974, B. Kendrick (KNZ 697).

Known distribution: New Zealand.
   C. curvata resembles *C. emodensis* and *C. scabrida*. It differs from *C. emodensis* in having larger phialides with longer, often curved, collarettes, and in its larger conidium bearing a minute marginal frill at its truncate base. It is distinguished

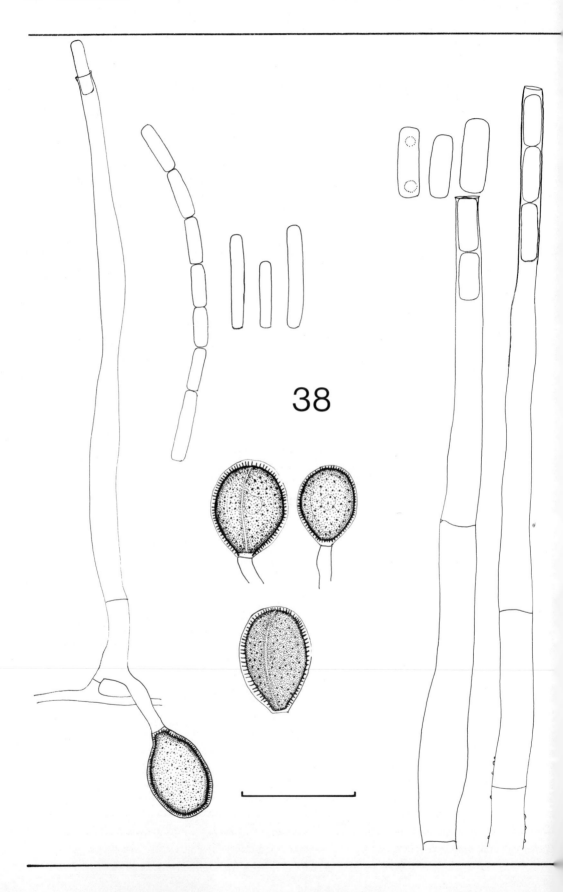

38

from *C. scabrida* by its predominantly lageniform phialides with abrupt transition from venter to collarette, and by its longer conidia.

21) Chalara cylindrica Karst. (Figure 29B)
in Meddn. Soc. Fauna Flora fenn., 14: 108, 1887.

Colony superficial, effuse, brown, hairy. Phialophores simple, cylindrical or somewhat irregular, erect or bent, few- to many-septate, dark brown; wall 1 μ thick, verrucose at the base and for most of the length, 25-150 [62] μ long and 3-6 [4.8] μ wide at the base, terminating in a phialide. Phialides lageniform, pale brown, 14-34.5 [23] μ long; venter usually ellipsoidal, 6.5-17 [13] x 4.5-6 [5] μ; collarette thin-walled, cylindrical, 6-15 [11] x 1.5-3 [2.2] μ; transition from venter to collarette abrupt; ratio of mean lengths of collarette and venter = 0.9:1. Phialoconidia extruded singly or in chains; cylindrical, both ends rounded, or apex rounded and base truncate, unicellular, hyaline, smooth-walled; 3-9.5 [5.3] x 1-1.5 [1.3] μ; mean conidium length/width ratio = 4:1.

Habitat: On *Abies, Eucalyptus, Picea.*

Specimens examined: 1) DAOM 41729, ex *Holotype* in Herb. Karsten, Bot. Mus. Helsinki, on *Abies*, Villias, 11.VI.1869; 2) IMI 69880, on *Picea*, VI.1869, authentic for *C. cylindrica*, ex Herb. Crypt. Mus. Paris; 3) IMI 82779, inside *Picea* bark, Salomonovice, Czechoslovakia, 31.VIII.1960, M. B. Ellis; 4) IMI 99717 ex Herb. Trail in Herb. Aberdeen Univ., (sub '*Chalara minuta* sp.n.') on rotten *Picea excelsa* needles, Persley Den, nr. Aberdeen, U.K., 22.III.1886; 5) FH 7109 ex C. Roumeguère—Fungi selecti exsiccati, on *Abies*, IV.1897, F. Fautrey; 6) BPI and B, sub *C. strobilina* on *Picea excelsa*, Siegen-Wald, Westfalen, 2.IV.1945, A. Ludwig; 7) B, ex Herb. Dr. A. Ludwig, Flora von Hessen-Nassau, sub *C. strobilina*, on *Picea excelsa*, Dillkreis, 16.IV.1938; 8) B, ex Herb. Dr. A. Ludwig, Flora von Westfalen, on *Picea excelsa*, Siegel, 2.IV.1938; 9) BR, ex Herb Bommer and Rousseau, on *Picea excelsa*, Fôret de St. Hubert, XI.1900; 10) DAOM 96204(d), on fallen bark of *Eucalyptus* sp., Alverstoke, Adelaide, S. Australia, 9.VII.1963, S. J. Hughes (1371d).

Known distribution: Australia, Czechoslovakia, France, Germany and U.K.
For affinities refer to *C. bohemica*.

22) Chalara cylindrosperma (Cda.) Hughes (Figure 35B)
in Can. J. Bot., 36: 747, 1958.
$\equiv$ *Menispora cylindrosperma* Cda.
  in Icon. Fung., 1: 16, 1837.
    = *Cylindrotrichum inflatum* Bon.
      in Handb. der Allgem. Mykologie, Stuttgart, p. 88, 1851.
    = *Chalara sanguinea* Höhn.
      in Sber. Akad. Wiss. Wien, Abt. I, 111: 63, 1902.
    = *Chalara flavopruinata* Höhn.
      in Mitt. bot. Inst. tech. Höchsch. Wien, 2:33, 1925.

Colony superficial, effuse, grayish brown, caespitose. Phialophores simple, densely crowded, erect or variously bent, cylindrical, septate; wall not constricted at septa, 1 μ thick and smooth; brown to dark brown in the basal part; 32-190 [88] μ long, basal cell slightly inflated and 3-7 [4.8] μ wide; terminating in a

---

Chalara state of *Ceratocystis radicicola*. Phialophores, conidia and chlamydospores ex IMI 51586.

39)   *Chalara elegans.* Phialophores, conidia and chlamydospores ex IMI 51546.

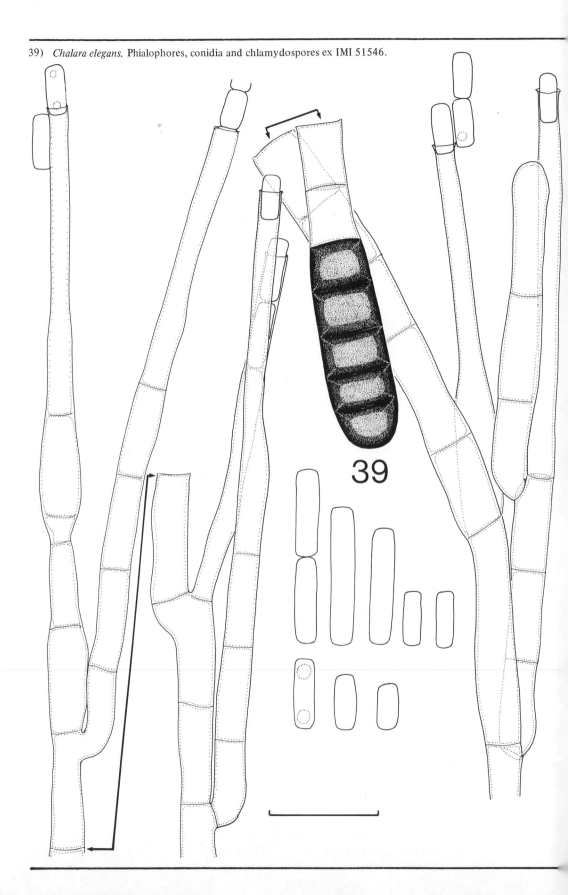

39

phialide. Phialides lagenform, light brown, 16-47 [32] μ long; venter ellipsoidal, 9.5-22 [15] x 4.5-8 [6] μ; collarette cylindrical, 12-28 [19] x 2-3 [2.7] μ; transition from venter to collarette abrupt; ratio of mean lengths of collarette and venter = 1.25:1. Proliferation rare, sympodial. Phialoconidia extruded singly or in easily dispersible or persistent chains; cylindrical, ends truncate or apex sometimes rounded; unicellular, hyaline, smooth; 5.5-17 [11] x 1.5-2.5 [1.9] μ; mean conidium length/width ratio = 6:1.

Habitat: On *Aconitum, Aesculus, Agathis australis, Betula alba, Fagus sylvatica, Gleditschia triacanthos, Ilex denticulata, Podocarpus dacrydioides, Podocarpus totara, Tilia* and unidentified wood.

Specimens examined: 1) IMI 44549 ex Herb. Corda in PR [*Holotype*] on rotting wood, nr. Reichenberg; 2) IMI 19219b, on *Fagus sylvatica* cupules on ground, Swinton Pk., Yorks., U.K., 11.X.1947, S.'J. Hughes; 3) IMI 54371 on *Aconitum rapelles*, Murthley Castle, Perthshire, U.K., IX.1953, M. B. Ellis; 4) IMI 90621b on *Aesculus* mast, Kingthorpe, Pickering, Yorks., U.K., 22.X.1961, W. G. Bramley; 5) IMI 120203g on *Ilex denticulata*, Gudalur, Nilgiris, India, 22.II.1966, K. A. Pirozynski (292g); 6) IMI 123698 on *Ilex denticulata*, Nilgiris, India, 22.II.1966, K.A.P. (32g); 7) FH 1624 in folder 11130 sub *Chalara sanguinea*, on decaying fruits of *Gleditschia triacanthos*, Austria; 8) FH 1618 in folder 11128, on *Tilia*, Germany, 6.X.1912, Feurlich (presumed type of *Chalara flavopruinata*); 9) on *Fagus* cupules, Waterloo, Ontario, Canada, 30.X.1968, T. R. Nag Raj; 10) WSP 57792 on decorticated branches, Witman Co., Pullman, Wash., U.S.A., 15.X.1967, P. R. Rao; 11) PC 2276, on fallen leaves of *Betula alba*, Bois de Gouards, S. of Versailles, France, 29-31.XI.1943, G. Arnaud; 12) PDD 32861 on dead leaves of *Agathis australis*, Unmarked Tr. off Scenic Dr., Waitakere Ra., Auckland, N.Z., 27.II.1974, B.K. (KNZ 420); 13) PDD 32864, on dead leaves of *Podocarpus dacrydioides*, Forest Hill Scenic Reserve, Southland Co., N. of Invercargill, N.Z., 18.IV.1974, B.K. (KNZ 659); 14) PDD 32862, on dead leaves of *Podocarpus dacrydioides*, Big Tree Tr., Peel Forest Pk., Canterbury Prov., N.Z., B.K. (KNZ 678); 15) PDD 32863, on dead leaves of *Podocarpus totara*, Peel Forest Pk., Canterbury Prov., N.Z., B.K. (KNZ 695).

Known distribution: Austria, Canada, Czechoslovakia, France, India, New Zealand, U.K.

C. cylindrosperma has overall resemblance to *C. longipes, C. nothofagi* and *C. stipitata.* It differs from *C. longipes* in possessing wider, ellipsoidal venters, longer collarettes and larger conidia. It can be distinguished from *C. stipitata* principally by its unicellular conidia and ellipsoidal rather than cylindrical or subcylindrical venter. It is distinct from *C. nothofagi* in its shorter phialides, ellipsoidal rather than subcylindrical venter, shorter and narrower collarettes, and narrower conidia without marginal frills at the base.

23) **Chalara dictyoseptata** Nag Raj & Hughes (Figure 12)
in N.Z. Jl. Bot., 12: 128, 1974.

Colony effuse, brown, hairy or tufted. Phialophores in lax clusters, 1-2-septate, fuliginous, 130-160 μ long, 8.5-10.5 μ wide at the base, terminating in a phialide. Phialides obclavate or lageniform, fuliginous, 120-160 [135] μ long, 14-20 [17] μ wide at the broadest part and 12-15 [14] μ wide at the apex of the collarette; wall asperate, transition from venter to collarette imperceptible. Phialoconidia extruded singly or in short chains; more or less cylindrical with a rounded apex and a truncate base showing prominent marginal frills; with up to 14 (usually 7) trans-

40

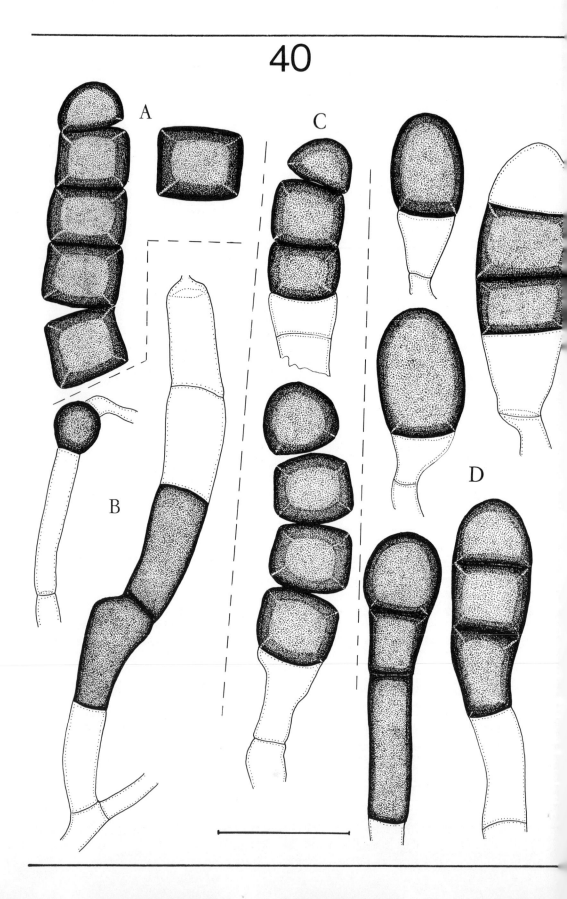

verse septa, 1-8 longitudinal septa; hyaline at first, later subhyaline to pale brown; wall smooth or irregularly roughened, not constricted at the septa; 32-62 [48] x 11-13 [12] μ; mean conidium length/width ratio = 4:1.

Habitat: On decayed stems of *Rhipogonum scandens.*

Specimen examined: DAOM 93336(a) ex *Holotype* in PDD 30407, Little Wanganui R., Westland, N.Z., 6.IV.1963.

Known distribution: New Zealand.

    C. dictyoseptata is a remarkable marginal species superficially resembling species of *Ascoconidium,* but disposed in *Chalara* because the walls of the phialides become progressively thinner towards the distal end of the deep collarette, and because there is a circumscissile split of the apex to release the conidia. The coloured, rough-walled dictyoconidia distinguish this species from all other known species of *Chalara.*

24) Chalara elegans sp. nov. (Figures 39, 40)
    Colonia effusa, primum albida et byssacea, deinde atrobrunnea et pulveracea. Phialophora simplicia, recta, cylindracea vel subcylindracea, 3-5-septata, nonnunquam aseptata, subhyalina vel pallide brunnea, pariete laevi, 70-95 μ long., in phialidem terminantia. Phialides lageniformes, interdum subcylindraceae vel obclavatae, 55-80 [69] μ long.; venter subcylindraceus, 20-40 [26] x 6-9.5 [7.2] μ; collum cylindraceum, 25-40 [33] x 3.5-5 [4.5] μ; transitio ex ventre ad collum gradatim vel abrupta; ratio long. colli et ventris = 1.2:1. Phialoconidia in catenas longas extrusa; cylindracea, interdum doliiformia, extremis truncatis vel obtusis, unicellularia, hyalina, subhyalina vel pallide brunnea, 7.5-19 (-30) [13] x 3-5 [4.2] μ, ratio conidii long./lat. = 3.1:1. Chlamydosporae thallicae, plerumque thallicae-endoarthricae, intercalares vel terminales, vulgo series rectilineares factae, ex 5-7 cellulis compositae, cujus apicalis chlamydospora conoidea, ceterae breves et cylindraceae, unicellulariae, atrobrunneae vel sucinaceae, 6.5-14 [10] x 9-13 [11] μ, pariete extimo tenui et laevi, etiam altero pariete intimo 1-2 μ cr.; rimae germinalium transversales, vulgo una ab utraque extremitate.

    Colony effuse, at first white and cottony, becoming brownish black and powdery. Phialophores simple, erect, cylindrical or subcylindrical, 3-5-septate or sometimes aseptate, subhyaline to pale brown, smooth-walled, 70-95 μ long, terminating in a phialide. Phialides lageniform, occasionally subcylindrical to obclavate, 55-80 [69] μ long; venter subcylindrical, 20-40 [26] x 6-9.5 [7.2] μ; collarette cylindrical, 25-40 [33] x 3.5-5 [4.5] μ; transition from venter to collarette gradual or occasionally abrupt; ratio of mean lengths of collarette and venter = 1.2:1. Phialoconidia extruded in long chains; cylindrical or occasionally doliiform with truncate or obtuse ends, unicellular, hyaline, subhyaline or pale brown, 7.5-19 (-30) [13] x 3-5 [4.2] μ; mean conidium length/width ratio = 3.1:1. Chlamydospores thallic, predominantly thallic-endoarthric, intercalary or terminal, usually in rectilinear series of 5-7, in which the terminal chlamydospore is conoid and the rest are short-cylindrical, unicellular, dark brown or amber, 6.5-14 [10] x 9-13 [11] μ with a thin, smooth outer wall and a secondary inner wall 1-2 μ thick; germ slits transverse, usually one at each end.

Chalara elegans. Chlamydospores. A. ex IMI 742, type of *Torula basicola;* B. ex IMI 19856; C. ex IMI 14192a; D. ex IMI 125845.

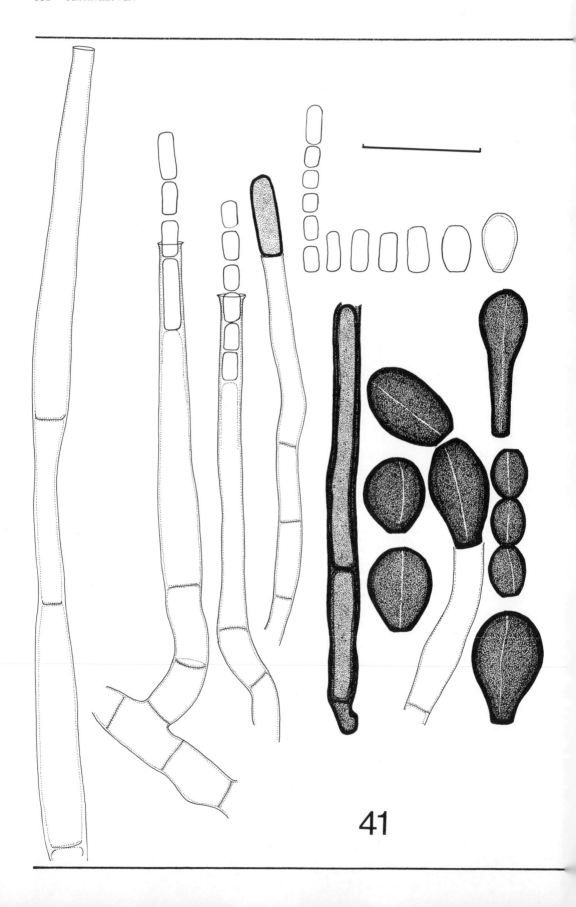

41

Habitat: On *Citrus, Crotalaria, Gloxinia, Lupinus, Lycopersicum, Nicotiana, Pisum, Primula,* and many other angiosperms, and in soil.

Specimens examined: 1) IMI 680, on seedling of *Nicotiana tabacum*, N.Z., Curtis; 2) IMI 742, Herb. Berkeley ex type coll. of *Torula basicola*, on stem bases of *Pisum*, 20.VI.1846, ex *Torula* folder in Herb. K; 3) IMI 14192(a), on roots of lemon seedlings, Cyprus, H. M. Morris; 4) IMI 19856, isol. ex *Citrus* soils, S. Calif., U.S.A., J. P. Martin; 5) IMI 21163, on tomato roots, Glamorgan, U.K., V.1944, S. J. Hughes; 6) IMI 21196, on *Primula* sp., Roy. Hort. Soc. Bot. Gdns., Wisley, U.K., 3.XII.1935, isol. D. E. Green; 7) IMI 35618, on *Crotalaria juncea*, Enterprise, S. Rhodesia; 8) IMI 51546, [*Holotype*] ex *Lupinus* sp., Queensland, Australia, R. Colbran; 9) IMI 125845, isol. ex *Citrus* soil, Israel, 12.IV.1965, R. Kenneth; 10) DAOM 108180 ex IMI 125777, on Belgian *Gloxinia* tubers recd. at Ottawa 20.IX.1963; 11) UAMH 1530, isol. ex soil, Ont., Canada, I.1963, G. L. Barron.

Known distribution: Widespread.

Cultural characters: MA at 28 °C in darkness, 48 hr.-old colony moist, white, aerial mycelium sparse, subsequently becoming dense and woolly. After 4-5 days, darkening through grayish white and gray, to brownish black; reverse dark gray; medium not discoloured.

Chalara elegans is unique among the fungi germane to this study in its thallic endo-arthric chlamydospores. It is the same fungus that has so far been known under the name *Thielaviopsis basicola* (Berk. & Br.) Ferr., of which *Torula basicola* Berk. & Br. is the basionym. The type specimen of *Torula basicola* bears, as has been demonstrated, only the thallic-arthric conidia, and under the Code the epithet *basicola* is restricted to this state only. Ferraris (1910), in transferring the epithet to *Thielaviopsis* as *T. basicola,* unwittingly made the two binomials obligate synonyms, despite providing a comprehensive description to include the phialidic state. In view of our concept that the phialidic state provides the generic name for the fungus in *Chalara,* and since the epithet *basicola* is not available, we are treating this taxon as an unnamed species of *Chalara.*

25) Chalara ellisii sp. nov. (Figure 29A)

Colonia superficialis, effusa, griseo-brunnea vel nigra, lanuginosa. Phialophora simplicia, cylindracea, septata, 32-185 μ long., 2.5-4.5 μ lat. ad basem, atrobrunnea et verrucosa ad basem, insuper pallescentia, in phialidem terminantia. Phialides subcylindraceae vel lageniformes, 20-38 [29] μ long.; venter cylindraceus vel subcylindraceus, 13-29 [21] x 3-4.5 [4] μ; collum obconicum vel subcylindraceum, 7-9.5 [8.4] x 2-2.5 [2.2] μ; transitio ex ventre ad collum abrupta; ratio long. colli et ventris = 0.4:1. Phialoconidia singulatim vel in catenas breves extrusa; cylindracea, extremis truncatis vel obtusis, unicellularia, hyalina, pariete laevi, 3-12 [6.3] x 1.5-2 [1.8] μ; ratio conidii long./lat. = 3.5:1.

Colony superficial, effuse, grayish brown or black, woolly. Phialophores simple, cylindrical, septate, 32-185 μ long, 2.5-4.5 μ wide at the base, dark brown and verrucose at the base, becoming paler above, terminating in a phialide. Phialides subcylindrical to lageniform, 20-38 [29] μ long, venter cylindrical or subcylindrical, 13-29 [21] x 3-4.5 [4] μ; collarette obconic or subcylindrical, 7-9.5 [8.4] x 2-2.5 [2.2] μ; transition from venter to collarette abrupt and marked by a constriction in the wall; ratio of mean lengths of collarette and venter = 0.4:1. Proliferation

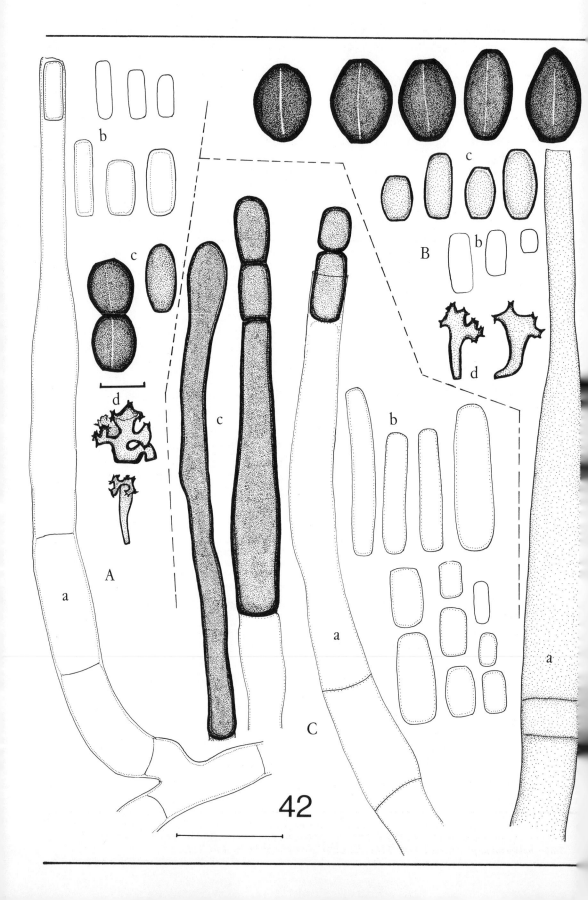

42

usually percurrent, occasionally sympodial. Phialoconidia extruded singly or in short chains; cylindrical with truncate or obtuse ends, unicellular, hyaline, smooth-walled, 3-12 [6.3] x 1.5-2 [1.8] μ; mean conidium length/width ratio = 3.5:1.

Habitat: soil.

Specimen examined: UAMH 1548 [*Holotype*], isol. ex soil, Guelph, Ontario, Canada, G. L. Barron (13601).

Known distribution: Canada.

Cultural characters: On MA (14.III.1969–29.III.1969) at 28 °C in darkness, phialophores subhyaline to pale brown, 38-103 μ long; phialides 20-38 μ long, venter 13-31 μ long, collarette 7.5-10 μ long, phialoconidia 4.5-13 x 1.5-2 μ.

   C. ellisii resembles *C. cylindrica,* but has subcylindrical phialides, rather obconic collarettes shorter than the venter, cylindrical or subcylindrical venters, and less verrucose walls.

26) Chalara emodensis sp. nov. (Figure 19B)

   Colonia obscura. Hyphae hyalinae vel subhyalinae, 2-2.5 μ lat., septata, ramosae, aggregatae, parietibus laevibus vel minute verrucosis, 0.5 μ cr., paucas et sparsas phialides ferentes. Phialides lageniformes vel ampulliformes, pallide brunneae, 16-28 [24] μ long.; venter ellipsoideus, 6.5-11 [8.7] x 7-11 [7.7] μ, collum cylindraceum, 15-18 [17] x 3-4 [3.7] μ; paries ventris vulgo verrucosus vel subtiliter verruculosus, paries colli laevis; transitio ex ventre ad collum abrupta; ratio long. colli et ventris = 1.9:1. Phialoconidia singulatim vel in catenas persistentes extrusa; cylindracea, extremis obtusis vel aliquantum rotundatis, 1-septata, hyalina, pariete laevi, 11-15 [13] x 2.5-3.5 [3] μ; ratio conidii long./lat. = 4.5:1.

   Colony obscure. Superficial mycelium composed of hyaline or subhyaline hyphae, 2-2.5 μ wide, septate at intervals of 5-9 μ, branched and aggregated; with smooth or minutely verrucose walls 0.5 μ thick; bearing scanty, scattered phialides. Phialides lageniform or ampulliform, pale brown, 16-28 [24] μ long; venter ellipsoidal, 6.5-11 [8.7] x 7-11 [7.7] μ; collarette cylindrical, 15-18 [17] x 3-4 [3.7] μ; venter wall usually rough or finely verrucose, collarette wall smooth; transition from venter to collarette abrupt; ratio of mean lengths of collarette and venter = 1.9:1. Proliferation not seen. Phialoconidia extruded singly or in persistent chains; cylindrical, ends blunt or rounded, 1-septate, hyaline, smooth-walled; 11-15 [13] x 2.5-3.5 [3] μ; mean conidium length/width ratio = 4.5:1.

Habitat: on leaves of *Quercus incana.*

Specimen examined: IMI 93970(b) [*Holotype*], Ghara Jahi, Pakistan, 4.IV.1962, S. Ahmad. (15550).

Known distribution: Pakistan.
   For affinities of *C. emodensis* refer to *C. curvata.*

27) Chalara fungorum (Sacc.) Sacc. (Figure 31C)
in Michelia, 1: 80, 1877.

Chalara paradoxa (conidial *Ceratocystis paradoxa*). a. Phialophores, b. conidia, c. chlamydospores, d. perithecial ornamentation. A. ex IMI 41297 (type of *Ceratocystis paradoxa*); B. ex holotype of *Stilbochalara dimorpha*; C. ex DAOM 75211 ex URM 640 (type of *Hughesiella euricoi*).

43A) *Chalara* state of *Ceratocystis moniliformis*. Phialophores and conidia ex IMI 125922.

43B) *Chalara ovoidea*. Phialophores, conidia and chlamydospores ex IMI 68863.

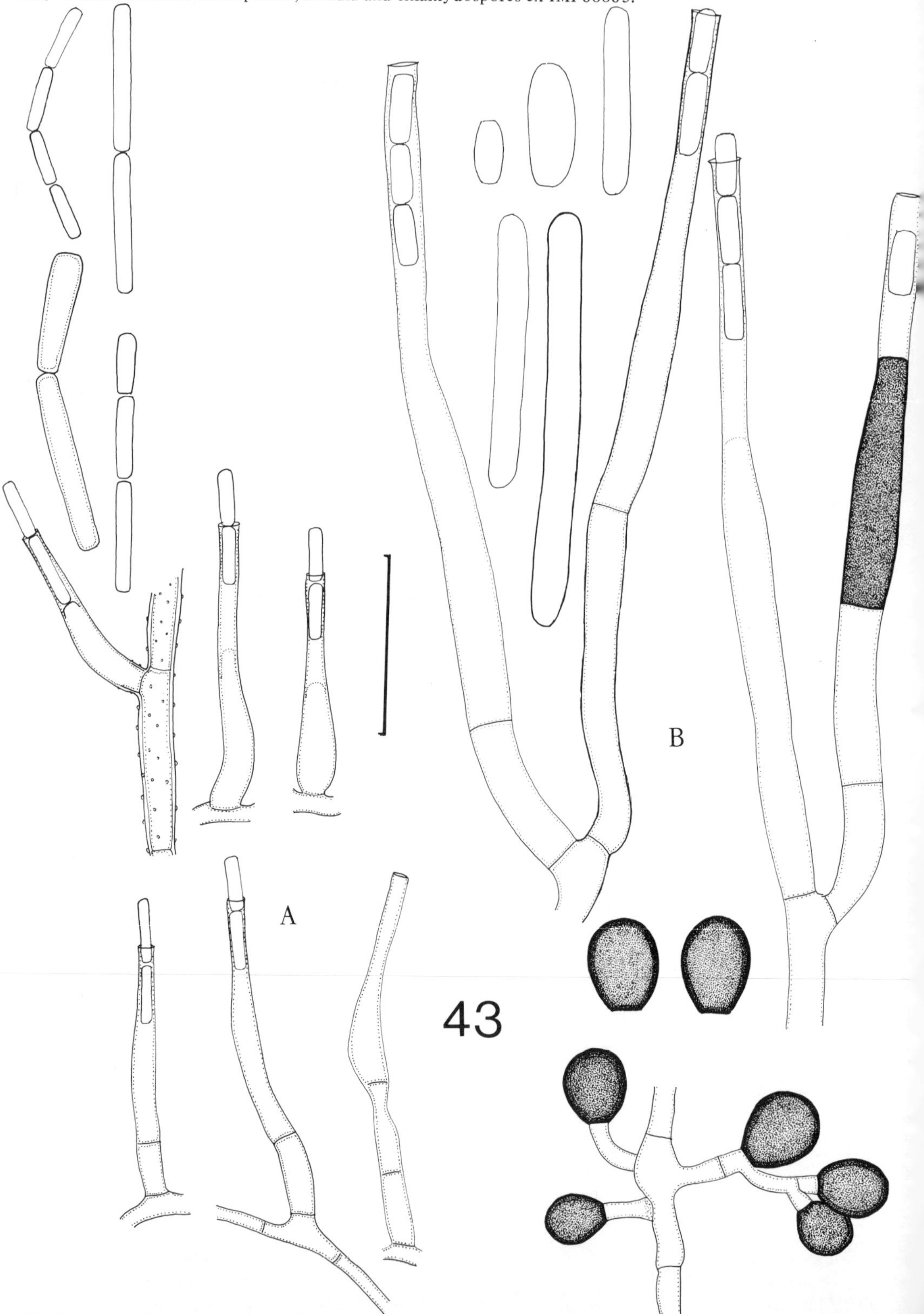

A

B

43

Chalara thielavioides. A. conidia and chlamydospores ex type in PAD 614.
Phialophores, conidia and chlamydospores ex DAOM 84174.

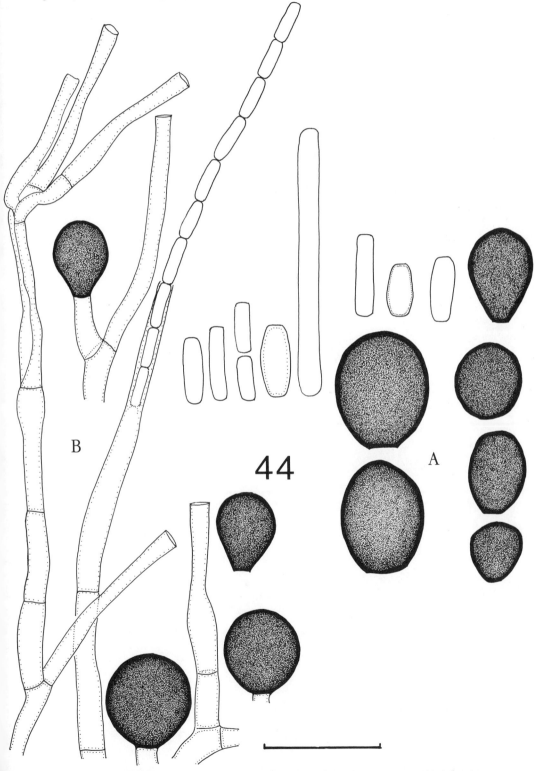

45) *Chalara* state of *Ceratocystis fimbriata.* Phialophores, conidia and chlamydospores ex IMI 28618.

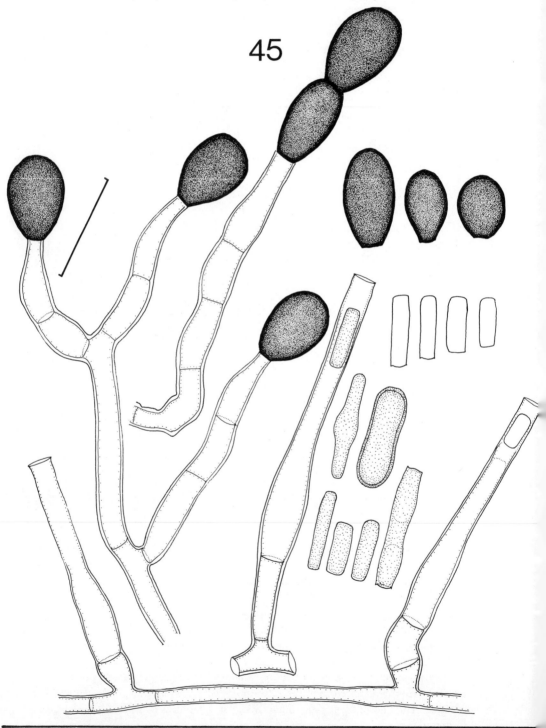

45

≡ *Cylindrium fungorum* Sacc.
in Atti Accad. scient. veneto-trent.-istriana, 2: 225, 1873.

Phialophores reduced to phialides, or simple with 1-2 indistinct septa. Phialides lageniform, brown to dark brown, paler above, smooth-walled; 25-40 [33] μ long, 3-6.5 [4.8] μ wide at the base; venter subcylindrical to ellipsoid, 12-19 [15] x 5.5-7.5 [6.4] μ; collarette cylindrical, 11-21 [17] x 2.5-4 [3.3] μ; transition from venter to collarette gradual; ratio of mean lengths of collarette and venter = 1.2:1; proliferation not seen. Phialoconidia cylindrical with flattened ends; unicellular, hyaline; 5.5-8 [7] x 2-2.5 [2.2] μ; mean conidium length/width ratio = 3.3:1.

Habitat: On *Hydnum compactum.*

Specimen examined: DAOM 43437 (slide) ex *Holotype* of *Cylindrium fungorum* Sacc. in PAD, XII.1872.

Known distribution: Italy
C. *fungorum* resembles *C. hughesii,* but has lageniform to somewhat irregularly shaped, smaller phialides, shorter and wider collarettes, gradual rather than abrupt transition from venter to collarette, and unicellular, shorter conidia.

28) Chalara fusidioides (Cda.) Rabenh. (Figure 30F)
in Kryptogamenflora, 1: 38, 1844.
≡ *Torula (Chalara) fusidioides* Cda.
in Icon. Fung., 2: 9, 1838.

Colonies barely perceptible, effuse, dirty white or yellowish white, pulverulent. Superficial hyphae 2.4-4 μ wide, subhyaline to pale brown, smooth-walled; bearing phialides directly or on short, 1-3-septate phialophores. Phialides lageniform, subhyaline to pale brown, smooth-walled, 11-26 [20] μ long; venter globose, rarely ellipsoidal, 5-11 [8] x 3.5-9 [6.2] μ; collarette cylindrical, 5.5-16 [12] x 1.5-4 [2.6] μ; transition from venter to collarette abrupt; ratio of mean lengths of collarette and venter = 1.56:1. Phialoconidia extruded singly or in easily dispersible chains; cylindrical with truncate, or occasionally rounded, ends, unicellular, hyaline, smooth-walled; 4.5-12 [7.7] x 1.5-3.5 [2.1] μ; mean conidium length/width ratio = 3.6:1.

Habitat: On *Fragaria vesca, Pinus, Podocarpus hallii, Vitis,* and old perithecia of *Mycosphaerella.*

Specimens examined: 1) PR 155685 [*Holotype*], *Pinus* bark; 2) PAV 1761, on rotting fronds and needles, nr. Hoyerswerda, E. Germany, Preuss, ex Rabenh. Klotz. Herb. viv. Myc.; 3) IMI 117139, on leaves of *Fragaria vesca,* Coel Shah, Nepal, India; 4) DAOM 88108, on old leaves of *Tofieldia pusilla,* Clyde inlet, Baffin Is., Franklin Dist., N.W.T., Canada (ex *Mycosphaerella minor* DAOM 63507), 31.VII.1950, P. Dansereau; 5) DAOM 70200, on old perithecia of *Mycosphaerella tassiana* on *Brya purpurascens,* Somerset Is., 72° 05' N, 94° 10' W, Franklin Dist. N.W.T., Canada, 22.VII.1958, D.B.O. Savile; 6) PDD 32866 on *Podocarpus hallii,* Governor's Bush Tr., Mt. Cook Nat. Pk., N.Z., 25.IV.1974,B.K. (KNZ 574).

Known distribution: Canada, Czechoslovakia, India, Italy, New Zealand.
Chalara fusidioides closely resembles *C. rhynchophiala* and, to a lesser degree, *C. emodensis* and *C. ampullula.* It differs from *C. rhynchophiala* in possessing smaller phialides with relatively short collarettes, and smaller conidia. It is distinguished from *C. emodensis* by its smooth-walled phialides and smaller, unicellular conidia. The morphology and dimensions of its phialides differentiate it from *C. ampullula.*

29) Chalara germanica Nag Raj & Kendrick (Figure 34A)
in Can. J. Bot., 49: 2121, 1971.

Colonies scattered, erumpent, dark brown, caespitose. Phialophores arising from stromatic aggregations of vegetative hyphae, densely crowded, simple, cylindrical, erect or variously bent, multiseptate, dark brown below, gradually becoming lighter above, smooth-walled, 62-156 [113] μ long and 4.5-7.5 [6.1] μ wide at the base, terminating in a phialide. Phialides lageniform, 33-42 [38] x 6-10 [8.6] μ; venter ovoid, conical, or subcylindrical, 16-20 (-24) [18] μ long; collarette subcylindrical, 16-23 (-25) [20] x 3.5-5 [4] μ; transition from venter to collarette abrupt or gradual; ratio of mean lengths of collarette and venter = 1.1:1. Phialoconidia extruded singly; cylindrical; apex rounded, base truncate with a minute marginal frill; 0-(rarely 1-)septate, hyaline, smooth-walled; 6.5-10 [8] x 2.5-3.5 [3.4] μ, mean conidium length/width ratio = 2.4:1.

Habitat: On twigs of an unknown tree.

Specimen examined: B250 mappe 230c [*Holotype*] (no other collection data).

Known distribution: Germany.
For affinities, refer to *C. brachyspora.*

30) Chalara ginkgonis Ferd. & Winge (Figure 36B)
in Bot. Tidsskr., 28: 256, 1907.

Colony forming irregular patches, effuse, brownish black, hairy. Phialophores solitary or in small clusters of 3-4, simple, cylindrical, up to 6-septate, sometimes slightly constricted at the septa, light brown at the base, yellowish brown above; wall 1 μ thick, smooth or verrucose; 48-93 [70] μ long, 4.5-5.5 [5] μ wide at the base; terminating in a phialide. Phialides lageniform to conical, yellowish brown, 29-56 [42] μ long; venter cylindrical or subcylindrical, 17-27 [21] x 4.5-8.5 [6.4] μ; collarette subcylindrical, 14-29 [21] x 2.5-3 [2.7] μ; transition from venter to collarette gradual; ratio of mean lengths of collarette and venter = 1:1. Proliferation of phialides not seen. Phialoconidia extruded singly or in easily dispersible short chains; cylindrical with blunt ends, usually 1-septate (sometimes unicellular), hyaline, smooth-walled and without constriction at the septa; 5-12 [8.6] x 2-3 [2.2] μ; mean conidium length/width ratio = 3.9:1.

Habitat: On leaves of *Ginkgo biloba.*

Specimen examined: C. [*Holotype*] on *Ginkgo biloba,* II.1907, Ferdinandsen and Winge.

Known distribution: Denmark.
C. ginkgonis is close to *C. affinis* but has usually multiseptate phialophores, longer phialides and shorter, narrower, usually 1-septate conidia.

31) Chalara gracilis sp. nov. (Figure 21B)

Colonia effusa, irregularis, brunnea, velutina, propter stratum albidum catenarum conidiorum pubescenti aspectu. Mycelium superficiale laxe aggregatum; hyphae noduliformes ob constrictiones parietis ad septa, brunneae et asperae, usque ad 4 μ lat. Phialophora ex segmentis hypharum orientia, brevi stipite 1-2-cellularum, vel plerumque ad sessiles phialides redacta; 40-52 μ long., infra 4-4.5 μ lat., praeter sufflatam asperam cellulam basalem, usque ad 7.5 μ lat. Phialides subcylindraceae; 33-44 [38.6] μ long., pallide brunneae, ad apicem pallescentes; venter subcylindraceus vel conicus, leniter bulbosus parte submediana, laevis, 13-18 [15.3] x 3.5-4.5 [4.1] μ; collum cylindraceum, laeve, 20-27 (-29) [23.3] x 2-2.5 μ; transitio ex ventre ad collum gradatim; ratio long. colli et ventris = 1.5:1. Phialoconidia singulatim vel in catenas breves extrusa, cylindracea; apice rotundato, base truncata, minutam fimbriam marginalem ferente, 1-septata, hyalina, laevia, 9-17 [14] x 1.5-2 [1.8] μ; ratio conidii long./lat. = 7.7:1.

Colony effuse, irregular, brown, hairy, with a downy aspect due to a whitish cover of chains of conidia. Superficial mycelium loosely aggregated; hyphae appearing nodulose due to constriction of the wall at septa, brown and asperate, up to 4 μ wide. Phialophores arising from segments of hyphae, with a short stalk of 1-2 cells, or often reduced to sessile phialides; 40-52 μ long, 4-4.5 μ wide below except for the swollen, asperate, basal cell which is up to 7.5 μ wide. Phialides subcylindrical, 33-44 [38.6] μ long, pale brown, becoming lighter at the apex; venter subcylindrical to conic, slightly bulbous in the submedian part, smooth-walled, 13-18 [15.3] x 3.5-4.5 [4.1] μ; collarette cylindrical, smooth, 20-27 (-29) [23.3] x 2-2.5 μ; transition from venter to collarette gradual; ratio of mean lengths of collarette/venter = 1.5:1. Phialoconidia extruded singly or in short chains; cylindrical, apex rounded, base truncate with minute marginal frill; 1-septate, hyaline, smooth; 9-17 [14] x 1.5-2 [1.8] μ; mean conidium length/width ratio = 7.7:1.

Habitat: On dead leaves of *Knightia excelsa.*

Specimen examined: PDD 32872 [*Holotype*] , Fairy Falls Tr., Waitakere Ra., Auckland, N.Z., 14.II.1974, B. Kendrick (KNZ 315).

C. gracilis is distinct from all other species of *Chalara* by virtue of its slender phialides, narrow venter that is bulbous in the submedian part; and narrow conidia.

32) Chalara hughesii Nag Raj & Kendrick (Figure 25B)
in Nag Raj & Hughes, N.Z. Jl. Bot. 12: 118, 1974.

Colony superficial, effuse, velutinous, brown, covered with white masses of conidia. Phialophores often reduced to phialides, but occasionally cylindrical, 1-2-septate, pale brown or brown, 33-54 [45] μ long, 4.5-5.5 [5] μ wide at the base; terminating in a phialide. Phialides ampulliform or lageniform, 25-48 [40] μ long; venter ellipsoidal or subcylindrical, 10-16 [14] x 5.5-7.5 [6.5] μ; collarette cylindrical, 19-29 [25] x 2-3 [2.6] μ; transition from venter to collarette abrupt; ratio of mean lengths of collarette and venter = 1.7:1. Proliferation not observed. Phialoconidia extruded singly or in chains; cylindrical with obtuse ends, 1-septate, hyaline, smooth-walled; 12-17 [14] x 2-2.5 [2.2] μ; mean conidium length/width ratio = 6.6:1.

Habitat: On *Hoheria angustifolia, Viburnum* sp., and undetermined wood.

Specimens examined: 1) DAOM 29354 [*Holotype*] , on undet. wood, Lloyd Cornell Preserve, Ringwood, N.Y., U.S.A., 6.IX.1952, S. J. Hughes; 2) DAOM 93533(b), on rotten wood, nr. Little R., Okitu Valley, Canterbury Dist., N.Z., 17.V.1963, S.J.H. (723b) (in PDD 21055); 3) DAOM 93848(b), on *Viburnum* sp., Epsom, Auckland, N.Z., 13.I.1963, S. Davison, S.J.H. (129b) (in PDD 20491); 4) DAOM 110037, on *Hoheria angustifolia*, nr. Little R., Okitu Valley, Canterbury Dist., N.Z., 17.V.1963, S.J.H. (750) (in PDD 21027); 5) DAOM 110038, on rotten wood, nr. Little R., Okitu Valley, Canterbury Dist., N.Z., 17.V.1963, S.J.H. (747) (in PDD 21030).

Known distribution: New Zealand, U.S.A.
For affinities refer to *C. fungorum.*

33) Chalara inaequalis sp. nov. (Figure 20A)

Colonia effusa, brunnea vel atrobrunnea, pubescens. Mycelium superficiale in stroma tenue ex cellulis atrobrunneis, usque ad 5 μ diam., laevibus, parietibus incrassatis compositum aggregatum. Phialophora ex stromate enascentia, solitaria vel aggregata, recta, 70-100 μ long. et usque ad 6.5 μ lat. ad basem; septis tenuibus 1-5; infra brunnea sed supra pallidiora, laevia, in phialidem terminantia. Phialides lageniformes, 54-65 [61] μ long., venter subcylindraceus vel ellipsoideus, 17-25 [20] μ long. et 5.5-8 [7.2] μ lat. parte supermediana; collum subcylindraceum, saepe versus apicem angustatum, fuscius quam venter, 29-47 [40] x 2.5-3.5 [2.9] μ; transitio ex ventre ad collum abrupta; ratio long. colli et ventris = 2:1. Phialoconidia catenulata, cylindracea, apice rotundato, base truncata fimbriam marginalem minutam ferente, 1-septata, cellula apicali breviore quam cellula basali; hyalina, laevia, 9.5-18 [13.5] x 2-2.5 μ; ratio conidii long./lat. = 5.8:1.

Colony effuse, brown to dark brown, hairy. Superficial mycelium aggregated to form a thin stroma of isodiametric, dark brown cells up to 5 μ diam., with smooth, thick walls. Phialophores arising from the stroma, solitary to aggregated, erect, 70-100 μ long and up to 6.5 μ wide at the base; with 1-5 thin septa; brown below but lighter above, smooth-walled, terminating in a phialide. Phialides lageniform, 54-65 [61] μ long, venter subcylindrical or ellipsoidal, 17-25 [20] μ long and 5.5-8 [7.2] μ wide at its supramedian part; collarette subcylindrical, often tapering toward the apex, darker than venter, 29-47 [40] x 2.5-3.5 [2.9] μ; transition from venter to collarette abrupt; ratio of mean lengths of collarette/venter = 2:1. Phialoconidia catenulate, cylindrical, apex rounded, base truncate with a minute marginal frill; 1-septate, apical cell shorter than basal cell; hyaline, smooth-walled, 9.5-18 [13.5] x 2-2.5 μ; mean conidium length/width ratio = 5.8:1.

Habitat: On dead leaves of *Nothofagus menziesii.*

Specimen examined: PDD 32643 [*Holotype*], Rd. to Milford Sound, 1 mi. N. of Tunnel, N.Z., 14.IV.1974, B. Kendrick (KNZ 676).

Known distribution: New Zealand.

C. inaequalis is distinct from all other known species of *Chalara* in its unequally 2-celled conidia. It is close to *C. cylindrosperma* in its phialophore morphology but can be distinguished by its longer phialides, longer collarettes, and its 1-septate conidia bearing basal marginal frills.

34) Chalara inflatipes (Pr.) Sacc. (Figure 13B)
in Sylloge Fung., 4:385, 1886; Nag Raj & Kendrick in Can.
J. Bot., 49: 2119, 1971.
≡ *Cylindrosporium inflatipes* Pr.
in Linnaea, 24: 106, 1851.
≡ *Chalara inflatipes* (Pr.) Lindau
in Rabenh. Kryptogamenflora, 8: 753, 1907.

Colony superficial, effuse, brownish black to black, setose to velutinous. Phialophores solitary, occasionally in groups of up to four, simple, cylindrical, 3-12-septate, brown to dark brown; wall verrucose, up to 1 μ thick and not constricted at the septa; 140-250 [180] μ long and 7.5-9.5 μ wide at the base; terminating in a phialide. Phialides subcylindrical to lageniform, (80-) 105-145 [126] μ long; venter subcylindrical, pale brown, 27-55 [41] x 9.5-15 [13] μ; collarette dark, cylindrical, (54-) 83-97 [89] x 5.5-8.5 [7] μ; transition from venter to collarette abrupt; ratio of mean lengths of collarette and venter = 2.1:1. Phialoconidia extruded singly or in easily dispersible, very short chains; cylindrical with a rounded apex and a truncate base having a well-defined marginal frill; usually 3-septate, occasionally variably septate, hyaline; wall smooth and not constricted at septa; 22-37 (-53) [31] x 4-5 [4.5] μ; mean conidium length/width ratio = 6.8:1.

Habitat: On *Alnus.*

Specimen examined: B. 228 (mappe 230b) [*Lectotype*] , between wood and bark of *Alnus* (no other collection data).

Known distribution: Germany.
    Affinities of this species are indicated under *C. bicolor.*

35) Chalara insignis (Sacc., Rouss., & Bomm.) Hughes (Figure 17A)
in Can. J. Bot., 36: 747, 1958.
$\equiv$ *Sporoschisma insigne* Sacc., Rouss., & Bomm.
    in Atti Ist. Veneto Sci., 6(2): 455, 1884.
    Colony superficial, effuse, brownish black to black, setose to velutinous. Phialophores solitary or in groups; simple, cylindrical, multiseptate, not constricted at the septa, brown to dark brown; wall verrucose and up to 1 μ thick; 70-220 [150] μ long and 8-13 [11] μ wide at the base; terminating in a phialide. Phialides subcylindrical to lageniform, 66-145 [110] μ long; venter light brown, subcylindrical, 28-61 [43] x 9.5-17 [13] μ; collarette dark brown, cylindrical, 40-115 [69] x 5.5-10 [8] μ; transition from venter to collarette abrupt; ratio of mean lengths of collarette and venter = 1.6:1. Phialoconidia extruded singly or in easily dispersible short chains; cylindrical with a rounded apex and a truncate base having a well-defined marginal frill; mostly 7-septate, occasionally variably septate, hyaline, smooth-walled and unconstricted at the septa; 18-54 [35] x 5-7 [6] μ; mean conidium length/width ratio = 5.8:1.

Habitat: On *Corylus avellana* and unidentified wood.

Specimens examined: 1) DAOM 28764 ex IMI 19052, on *Corylus avellana* periderm and wood, Steynbridge, Devon, U.K., 15.X.1947, S. J. Hughes; 2) DAOM 29422 on old *Poria* and wood, Lloyd Cornell Preserve, McLean, N.Y., U.S.A., 5.IX.1952, S.J.H.; 3) DAOM 51011 (slide) ex Herb. Saccardo in PAD [*Holotype*] S.143 sub *Sporoschisma insigne.*
    C. insignis resembles *C. inflatipes* and *C. bicolor*; in fact, it has been regarded as a synonym of *C. inflatipes* (Nag Raj & Kendrick 1971). Dr. S. J. Hughes (pers. comm.) considers that before such a synonymy could be accepted, the gap between these two species would need to be bridged by more collections than are available at the moment. Accordingly, we are persuaded to maintain them as distinct.
    Affinities of *C. insignis* are indicated under *C. bicolor.*

36) Chalara kendrickii Nag Raj sp. nov. (Figure 21C)
    Colonia superficialis, effusa, brunnea ad atrobrunnea, velutina. Phialophora ex tenui stromate cellularum cubicarum vel irregularium, incrassatarum atrobrunnearum orientia plerumque ad phialides redacta; vel simplicia, cylindracea, usque ad 2-septata, 41-72 μ long., 5-8 μ lat. ad basem, flavide brunnea, pallide brunnea vel brunnea, in phialidem terminantia. Phialides lageniformes vel obclavatae, 41-67 [53] μ long.; venter subcylindraceus vel ellipsoideus, 10-28 [18] x 4-11 [7.3] μ; collum cylindraceum, 18-40 [30] x 2.5-6 [4.2] μ; transitio ex ventre ad collum gradatim; ratio long. colli et ventris = 1.6:1. Phialoconidia singulatim vel in catenas breves extrusa; cylindracea, apice rotundato et base truncata cum fimbria margin-

alem ferente, 1-septata, hyalina, pariete laevi, 8-13 [12] x 2-4 [3] μ; ratio conidii long./lat. = 3.7:1.

Colony superficial, effuse, brown to dark brown, hairy. Phialophores arising from a thin stromatic layer of thick-walled, dark brown, cuboid or irregularly shaped cells; often reduced to phialides, or simple, cylindrical, up to 2-septate; 41-72 μ long, 5-8 μ wide at the base; yellowish brown, pale brown or brown; terminating in a phialide. Phialides lageniform or obclavate, 41-67 [53] μ long; venter subcylindrical or ellipsoidal, 10-28 [18] x 4-11 [7.3] μ; collarette cylindri-cal, 18-40 [30] x 2.5-6 [4.2] μ; transition from venter to collarette gradual; ratio of mean lengths of collarette and venter = 1.6:1. Proliferation not observed. Phialo-conidia extruded singly or in short chains; cylindrical with a rounded apex and a truncate base bearing a marginal frill; 1-septate, hyaline, smooth-walled; 8-13 [12] x 2-4 [3] μ; mean conidium length/width ratio = 3.7:1.

Habitat: On undetermined species of *Baissea* and *Rubus.*

Specimens examined: 1) IMI 17309(c), on dead stems of *Rubus,* Cwm woods, Aberystwyth, U.K., 5.VIII.1947; 2) IMI 58097 (no collection data); 3) IMI 61712(a) [*Holotype*], on leaves of *Baissea* sp., Njala (Kori), Sierra Leone, 26.IX.1955, C. T. Pyne (M6364).

Known distribution: Sierra Leone, U.K.

Affinities of *C. kendrickii* are indicated under *C. angionacea.*

IMI 58097 bears inadequate material of another undetermined species of *Cha-lara* with phialides similar to those of *C. kendrickii* (Figure 21D), but smaller (25-29 μ long and 4-5 μ wide at the broadest point on the venter and 2.5 μ wide at the apex), and some phialoconidia that are unicellular and 3-4 x 2 μ.

37) Chalara longipes (Pr.) Cooke (Figure 33B)
in Grevillea, 10: 50, 1881; Sacc., Sylloge Fung., 4: 335, 1886; Nag Raj & Kendrick, in Can. J. Bot., 49: 2120, 1971.
≡ *Cylindrosporium longipes* Pr.
in Linnaea, 24: 106, 1851.
≡ *Chalara longipes* (Pr.) Lindau
in Rabenh. Kryptogamenflora, 8: 752, 1907.

Colony inconspicuous, superficial, effuse, brown. Phialophores densely crowded, simple, cylindrical, multi-septate, septa in distal part indistinct; brown to dark brown below, lighter above; wall smooth and not constricted at the septa; 63-120 μ long, slightly swollen at the base (3.5-4 μ wide), terminating in a phialide. Phialophores occasionally proliferate sympodially 1-3 times from below the phi-alide. Phialides lageniform, 21-29 [25] μ long; venter subcylindrical, 15-18 (-20) [17] x 3-5 [3.9] μ; collarette cylindrical or obconic, 5.5-9.5 [8.4] x 1.5-2 [1.6] μ; transition from venter to collarette gradual; ratio of mean lengths of collarette and venter = 0.5:1. Phialoconidia extruded in easily dispersible chains; cylindrical with obtuse or truncate ends, unicellular, hyaline, smooth-walled; 3.5-6.5[5] x 1-1.5 [1.2] μ; mean conidium length/width ratio = 4:1.

Habitat: On needles of *Pinus.*

Specimen examined: B. 227 (mappe 230b) [*Lectotype*] (no other collection data).

Known distribution: Germany.

Affinities of this species are indicated under *C. cylindrosperma.*

38) Chalara microspora (Cda.) Hughes (Figure 30C)
in Can. J. Bot., 36: 747, 1958.
≡ *Fusidium clandestinum β microsporum* Cda.
   in Icon. Fung., 2: 43, 1838.
≡ *Cylindrium clandestinum* (Cda.) Sacc. var. *microsporum* (Cda.) Sacc.
   in Sylloge Fung., 4: 37, 1886.

Colony superficial, effuse, dull white, downy. Phialophores scattered to gregarious, simple, erect, cylindrical, mostly 2- (sometimes 4-) septate, pale brown, 19-51 [30] μ long and 2.5-6.5 [3.5] μ wide at the base; walls smooth, 1 μ thick and moderately constricted at the septa; terminating in a phialide, or often reduced to phialides borne directly on vegetative hyphae. Phialides obclavate to lageniform, 18-36 [23] μ long; venter subcylindrical, 10-16 [14] x 2.5-6.5 [3.9] μ; collarette cylindrical, 6-17 [11] x 1-2.5 [1.7] μ; transition from venter to collarette abrupt, or rarely gradual; ratio of mean lengths of collarette and venter = 0.9:1. Proliferation rare, more often percurrent than sympodial. Phialoconidia extruded in long, persistent, or easily dispersible chains; cylindrical with blunt or truncate ends; unicellular, hyaline, smooth-walled; 3-8.5 [5.4] x 1-1.5 [1.1] μ, mean conidium length/width ratio = 5:1.

Habitat: On a member of Hydnaceae, *Quercus.*
Specimens examined: 1) DAOM 51799 (slide) ex Herb. Corda in PR 155502 [*Holotype*], sub '*Fusidium clandestinum β microsporum* Ca. Tuch. *querc.* ex Co-Type collection of *Fusidium clandestinum microsporum* Cda.' (C. 75); 2) FH 158, on hydnaceous fungus, Thaxter estate, Kittery Pt., Maine, U.S.A., XI.1939, D. H. Linder, sub *C. minima* v. Höhn. vel aff.

Known distribution: Czechoslovakia, U.S.A.

The Farlow specimen bears somewhat larger phialides, with slightly longer collarettes and wider conidia than are found in the type, but is otherwise almost identical with it. Rather than establish a new species on the basis of these minor differences, we prefer to broaden the existing species concept slightly to include this specimen.

Affinities of this species are considered under *C. austriaca.*

39) Chalara nigricollis sp. nov. (Figure 20B)
Colonia superficialis, effusa, brunnea, velutina. Phialophora solitaria, sparsa, interdum aggregata, recta vel flexa, cylindracea, usque ad 3-septata, pallide brunnea, usque ad 87 μ long., 3.5-9 [6.5] μ lat. ad basem; in phialidem terminantia, vel plerumque ad phialides redacta. Phialides lageniformes, 47-87 [65] μ long.; venter pallide-brunneus, 11-38 [26] x 5-13 [9] μ; collum cylindraceum, atrobrunneum, 32-48 [41] μ long. et 2.5-5 [3.7] μ lat.; transitio ex ventro ad collum abrupta; ratio long. colli et ventris = 1.5:1. Phialoconidia singulatim vel in catenas facile dissilientes extrusa; cylindracea, utrinque truncata, 1-septata, hyalina, laevia, 8.5-17 [13] x 2-2.5 [2.4] μ; ratio conidii long./lat. = 5.3:1.

Colony superficial, effuse, brown, hairy. Phialophores solitary and scattered, sometimes gregarious, erect or bent, cylindrical and up to 3-septate, pale brown; up to 87 μ long, 3.5-9 [6.5] μ wide at the base; terminating in a phialide; or often reduced to a phialide. Phialides lageniform, 47-87 [65] μ long; venter pale brown, 11-38 [26] x 5-13 [9] μ; collarette cylindrical, dark brown, 32-48 [41] x 2.5-5 [3.7] μ; transition from venter to collarette abrupt; ratio of mean lengths of collarette and venter = 1.5:1. Proliferation percurrent or sympodial. Phialoconidia extruded singly or forming easily dispersible short chains; cylindrical with truncate ends; 1-septate, hyaline, smooth-walled; 8.5-17 [13] x 2-2.5 [2.4] μ; mean conidium length/width ratio = 5.3:1.

Habitat: On *Cyperus longus.*

Specimens examined: IMI 55278 [*Holotype*] and IMI 58096, Petit Bot , Guernsey, U.K., 16.IV.1947, E. A. and M. B. Ellis.

Known distribution: U.K.

In possessing a collarette that is uniformly darker than the venter, *Chalara nigricollis* shows an affinity with *C. inaequalis,* from which it differs in its usually reduced phialophores, and medially septate conidia without a basal marginal frill.

40) Chalara nothofagi sp. nov. (Figure 35A)

Colonia effusa, brunnea vel atrobrunnea, velutina. Hyphae superficiales pallide brunneae, laeves, tenuiter aggregatae. Phialophora solitaria vel laxe conferta, recta vel varia flexa, 115-168 μ long., 5-6 μ lat. ad basem, cellula basali parum inflata, subglobosa et usque ad 7.5 μ diam.; multiseptata, infra brunnea vel atrobrunnea, supra pallescentia, in phialidem terminantia. Phialides lageniformes, 48-58 [53] μ long., pallide brunneae; venter subcylindraceus, 20-27 [23.6] x 6-8.5 [7.4] μ; collum cylindraceum, 26-33 [29] x 3.5-4 μ; transitio ex ventre ad collum abrupta; ratio long. colli et ventris = 1.2:1. Phialoconidia singulatim vel in catenas breves extrusa; cylindracea, apice rotundato, base truncata, minutam fimbriam marginalem ferente; unicellularia, hyalina, laevia, 13-17 (-18) [15] x 2.5-3 [2.7] μ; ratio conidii long./lat. = 5.5:1.

Colony effuse, brown to dark brown, hairy. Superficial hyphae thinly aggregated, pale brown, smooth. Phialophores solitary or in lax clusters, erect or variously bent; 115-168 μ long, 5-6 μ wide at the base; basal cell slightly inflated, subglobose and up to 7.5 μ diam.; multiseptate, brown or dark brown below, becoming lighter above; with a terminal phialide. Phialides lageniform, 48-58 [53] μ long, pale brown; venter subcylindrical, 20-27 [23.6] x 6-8.5 [7.4] μ; collarette cylindrical, 26-33 [29] x 3.5-4 μ; transition from venter to collarette abrupt; ratio of mean lengths of collarette/venter = 1.2:1. Phialoconidia extruded singly or in short chains; cylindrical, apex rounded, base truncate with a minute marginal frill; unicellular, hyaline, smooth; 13-17 (-18) [15] x 2.5-3 [2.7] μ; mean conidium length/ width ratio = 5.5:1.

Habitat: On dead leaves of *Nothofagus solandri* var. *cliffortioides.*

Specimen examined: PDD 32847 [*Holotype*] , Scott's Tr., Arthur's Pass Nat. Pk., N.Z., 27.IV.1974, B. Kendrick (KNZ 715).

Known distribution: New Zealand.

For affinities see *C. cylindrosperma.*

41) Chalara novae-zelandiae sp. nov. (Figure 33C)

Colonia effusa, brunnea vel atrobrunnea, velutina. Phialophora dense conferta, recta vel varie flexa, 45-120 μ long., 3-4.5 μ lat. ad basem; cellula basalis parum inflata, subglobosa vel conica, et usque ad 6.5 μ lat.; multiseptata, pariete laevi et aliquantum contracto ad septa; infra brunnea vel atrobrunnea, supra pallescentia, in phialidem terminantia. Phialides lageniformes, pallide brunneae, laeviae, 21-34 [28] μ long.; venter subcylindraceus, 11-19 [15] x 3.5-4.5 [4] μ; collum cylindraceum, 10-15 [12.7] x 1.5-2 μ; transitio ex ventre ad collum abrupta, raro gradatim; ratio long. colli et ventris = 0.8:1. Phialoconidia in catenas longas extrusa; cylindracea, apice rotundato, base truncata, minutam fimbriam marginalem ferente; unicellularia, hyalina, laevia; 5-8 [6.4] x 1-1.5 μ; ratio conidii long./lat. = 5:1.

Colony effuse, brown or dark brown, hairy. Phialophores densely crowded, erect or variously bent; 45-120 μ long, 3-4.5 μ wide at the base; basal cell slightly swollen, subglobose to conical and up to 6.5 μ wide; multiseptate, wall smooth and somewhat constricted at the septa which are thinner in the distal part; brown to dark brown below, becoming lighter above; with a terminal phialide; lateral proliferation of phialophore common. Phialides lageniform, pale brown, smooth-walled, 21-34 [28] μ long; venter subcylindrical, 11-19 [15] x 3.5-4.5 [4] μ; collarette cylindrical, 10-15 [12.7] x 1.5-2 μ;transition from venter to collarette abrupt, rarely gradual; ratio of mean lengths of collarette/venter = 0.8:1. Phialoconidia extruded in long chains; cylindrical, apex rounded, base truncate with a minute marginal frill; unicellular, hyaline, smooth; 5-8 [6.4] x 1-1.5 μ; mean conidium length/width ratio = 5:1.

Habitat: On dead leaves of *Nothofagus menziesii* and *Podocarpus dacrydioides.*

Specimens examined: 1) PDD 32848 [*Holotype*], on leaf of *N. menziesii,* Governor's Bush Tr., Mt. Cook Nat. Pk., N.Z., 25.IV.1974, B. Kendrick (KNZ 583b); 2) PDD 32849, on dead leaf of *P. dacrydioides,* Big Tree Tr., Peel Forest Pk., Canterbury Prov., N.Z., 26.IV.1974, B.K. (KNZ 682b).

Known distribution: New Zealand.

C. novea-zelandiae resembles *C. longipes,* but has proportionately shorter and narrower phialides, narrower venters, longer collarettes, and abrupt transition. It also exhibits affinities with *C. parvispora,* but can be distinguished by its shorter and narrower phialophores, shorter and narrower phialides, shorter and narrower collarettes and longer, narrower conidia.

42) Chalara ovoidea sp. nov. (Figure 43B)

Colonia superficialis, effusa, atra, pulveracea. Phialophora simplicia, cylindracea, hyalina, septata, 80-105 μ long., 4.5-5 μ lat. ad basem; in phialidem terminantia. Phialides subcylindraceae, 45-51 [48] μ long.; venter subcylindraceus, 19-37 [28] x 5-6.5 [6] μ; collum cylindraceum, 23-46 [34] x 2.5-4.5 [3.3] μ; transitio ex ventre ad collum gradatim vel vix perspicua; ratio long. colli et ventris = 1.2:1. Phialoconidia singulatim vel in catenas formantia; cylindracea, extremis rotundatis vel obtusis, unicellularia, hyalina, subhyalina vel brunnea, pariete laevi vel subtiliter verrucoso; 5-22 (-25) [12] x 2.5-5 (-6) [3.6] μ; ratio conidii long./lat. = 3.3:1. Chlamydosporae plerumque holoblasticae, subglobosae, pyriformes, basibus truncatis, unicellulariae, brunneae ad atrobrunneae; parietibus laevibus vel subtiliter verrucosis, 1.5 μ cr.; 7.5-14 (-22) [10] x 6-11 (-19) [8] μ; solitariae et terminales ex hyalinis, cylindraceis, sympodialiter ramificantibus conidiophoris orientes; rima germinalium verticalis vel obscura.

Colony superficial, effuse, black, powdery. Phialophores simple, cylindrical, hyaline, septate, 80-150 μ long., 4.5-5 μ wide at the base; terminating in a phialide. Phialides subcylindrical, 45-51 [48] μ long; venter subcylindrical, 19-37 [28] x 5-6.5 [6] μ; collarette cylindrical, 23-46 [34] x 2.5-4.5 [3.3] μ; transition from venter to collarette gradual or almost imperceptible; ratio of mean lengths of collarette and venter = 1.2:1. Phialoconidia extruded singly or forming chains; cylindrical with rounded or blunt ends, unicellular, hyaline, subhyaline, or brown; wall smooth or finely verrucose; 5-22 (-25) [12] x 2.5-5 (-6) [3.6] μ; mean conidium length/ width ratio = 3.3:1. Chlamydospores predominantly holoblastic; subglobose or pyriform with a truncate base, unicellular, brown to dark brown; walls smooth or minutely verrucose, 1.5 μ thick; 7.5-14 (-22) [10] x 6-11 (-19) [8] μ; solitary and terminal on sympodially branching, hyaline, cylindrical conidiophores; germ slit vertical or obscure.

Habitat: On *Fagus sylvatica.*

Specimens examined: 1) IMI 33730(a), on *Fagus sylvatica* wood, Ranmore, Surrey, U.K., 23.I.1949, S. J. Hughes; 2) IMI 34449, on *Fagus sylvatica* wood, Perranzabuloe, Cornwall, U.K., II.1949, F. Rilstone; 3) IMI 30934(a), on *Fagus sylvatica* floor board, Denmark, E. Jørgenson; 4) IMI 50910 [*Holotype*] , ex *Fagus sylvatica* wood, Denmark, 14.VI.1950, E.J.; 5) IMI 50911, ex *Fagus syvatica* (railway sleeper/tie), Seeland, Denmark, 25.VIII.1950, E.J.; 6) IMI 50665, isol. ex *Fagus sylvatica,* Knowles Pk., Sevenoaks, Kent, U.K., VIII.1952, M. B. Vincent; 7) IMI 50912, isol. ex *Fagus sylvatica,* Svendborg, Denmark, 9.XI.1950, E.J.; 8) IMI 68863, ex *Fagus* planks, Oxford, U.K., 22.III.1056, J. G. Savory.

Known distribution: Denmark, U.K.

The affinities of this species are indicated under *Chalara thielavioides.*

43) Chalara paradoxa (de Seynes) Sacc (Figures 41, 42)
in Sylloge Fung., 10: 595, 1892.

≡ *Sporoschisma paradoxum* de Seynes
   in Recherches pour servir à d'Histoire naturelle des Végétaux inférieurs, 3: 30, 1886.

≡ *Thielaviopsis paradoxa* (de Seynes) Höhn. in Hedwigia, 43: 295, 1904.

   = *Thielaviopsis ethacetica* Went. in Archiv, voor de Java Suekerrind., 1893, p. 4.
   = *Endoconidium fragrans* Delacr. in Bull. Soc. mycol. Fr. 9: 184, 1893, fide Höhnel, in Sber. Akad. Wiss. Wien, Abt. I, 118: 168, 1909.
   = *Stilbochalara dimorpha* Ferd. & Winge in Bot. Tidsskr. 30: 220, 1910.
   = *Hughesiella euricoi* Bat. & Vital in Anais. Soc. Biol. Pernamb., 14: 42, 1956.

Colony effuse, downy to cottony, becoming waxy in irregular patches, dirty white or yellowish white. Vegetative hyphae immersed in host tissue, up to 7.5 μ wide, hyaline, subhyaline or pale brown. Phialophores more or less hyphoid, simple, cylindrical to subcylindrical, up to 3-septate, hyaline or subhyaline, 55-125 μ long, 3-7.5 μ wide at the base; terminating in a phialide. Phialides lageniform or subcylindrical, 45-110 [80] μ long; venter cylindrical or subcylindrical, 4-12 (-14) [7.7] μ wide; collarette cylindrical, 2.5-6.5 [4.4] μ wide; transition from venter to collarette gradual or almost imperceptible. Phialoconidia extruded in long chains; cylindrical or doliiform, with rounded, blunt or truncate ends; unicellular, hyaline, subhyaline, or pale brown, smooth-walled; 3-25 [10] x 2.5-4.5 [3.5] μ; mean conidium length/width ratio = 2.8:1. Chlamydospores mostly enteroblastic-phialidic, less often holoblastic; ovoid, doliiform, ellipsoidal, or pyriform; unicellular, brown to reddish brown with smooth or minutely verrucose walls up to 1.5 μ thick; 9.5-25 [16] x 7-15 [11] μ; occurring in long terminal chains, or in clusters, or singly; germ slit vertical; frequently occurring in the form of thallic, irregular, sparsely-septate, hyphal elements.

Habitat: Ananas, Cocos, Elaeis, Ipomoea, Musa, Theobroma, etc.

Specimens examined: 1) IMI 15035, isol. ex *Musa* sp., Jamaica, VIII.1929; 2) IMI 15302, ex Herb. Mycol. M. C. Cooke 1885 sub *Torula (herbarum ?) Oidium chartarum* [Scr. M. C . Cooke] , on *Ananas,* VII. 1874, M. de Seynes (ex *Torula herbarum* folder in K.); 3) IMI 39075, isol. ex endosperm of *Cocos nucifera,* London, 1950, comm. G. Smith (BB 246); 4) IMI 41297 (*Holotype* of *Ceratocystis paradoxa*), on *Theobroma cacao,* saprobic on discarded husks, Ghana, 1927, H. A. Dade (CB 449); 5) IMI 52042, isol. ex *Musa* sp., Tanzania, comm. 7.IV.1953, G. B. Wallace; 6) IMI 53190, isol. ex *Elaeis guineensis,* Onitsha Prov., Nigeria, 23.I.1952, R. A. Bull; 7) IMI 62475, isol. ex *Ananas comosus,* Malaysia, A. Johnston; 8) IMI 62603, isol. ex *Ipomoea batatas* tubers, 11.IV.1956, comm. A. J.; 9) IMI 79766, ex *Elaeis guineensis,* Nigeria, 5.II.1960, J. S. Robertson; 10) IMI 92887, isol. ex *Cocos nucifera* leaves, Ceylon [Sri Lanka], 25.IV.1962; 11) IMI 94847, isol. ex *Cocos nucifera* leaves, Ceylon [Sri Lanka], 3.VIII.1962; 12) IMI 96377, isol. ex *Ananas comosus,* Kenya, 14.IX.1962, J. J. Njorge; 13) IMI 110880, isol. ex *Cocos nucifera,* Ogidi, E. Nigeria, 31.XII.1964, I. G. Weststeijn; 14) DAOM 74680, on *Ananas comosus* from Calif., U.S.A., G. L. Hennebert #1869; 15) DAOM 89810, on *Musa* sp.,Ottawa, Ont., Canada, 4.X.1962, R. E. Scharfe; 16) C (*Holotype* of *Stilbochalara dimorpha*) in Mus. Bot. Hauniense, on rotting pericarp of *Theobroma cacao*, Las Trincheras, Venezuela, 25.XII.1891, H. Lassen; 17) DAOM 75211 ex URM 640 (*Holotype* of *Hughesiella euricoi*), isol. ex air, San Salvador, Eurico de Matta.

Known distribution: Widespread.

Typification of Chalara paradoxa: Sporoschisma paradoxum de Seynes is the basionym of *Chalara paradoxa*. De Seynes did not designate the holotype; he indicated neither the origin of the specimen nor the herbarium in which it was deposited. Our attempts to trace his specimens have not been successful, and there is little reason to expect that the type still exists. De Seynes published illustrations of the fungus which form part of the protologue of the name and furnish adequate information for identification. These illustrations are, therefore, here designated as the type of the name.

Chalara paradoxa is separable from all other *Chalara* species with chlamydospores by its robust phialides, its indiscriminate formation of chlamydospores— thallic, enteroblastic-phialidic, holoblastic—and in the morphology of the phialides, phialoconidia and chlamydospores.

44) **Chalara parvispora** Nag Raj & Hughes (Figure 34C)
in N.Z. Jl. Bot., 12: 117, 1974.

Colony superficial, effuse, brown, hairy, covered with a thin white bloom of conidia. Phialophores solitary and scattered, erect or variously bent, dark brown below, gradually becoming lighter above; multi-septate; wall smooth and not constricted at the septa; 95-190 μ long, inflated at the base to a width of up to 6 μ, then 3-5 μ wide above; terminating in a phialide. Phialides lageniform, pale brown and smooth-walled; 34-43 [40] μ long; venter subcylindrical, 15-20 [17] x 4-7 [5.2] μ; collarette cylindrical, 18-26 [22] x 1.5-2.5 [2.3] μ; transition from venter to collarette abrupt, rarely gradual; ratio of mean lengths of collarette and venter = 1.3:1. Phialoconidia extruded singly or in persistent chains; short-cylindrical, with a rounded apex and a truncate base bearing a minute marginal frill; unicellular, hyaline, smooth-walled; 3.5-6 [4.7] x 1.5-2 [1.8] μ; mean conidium length/width ratio = 2.6:1.

Habitat: On *Cyathea medullaris.*

Specimen examined: DAOM 110036 ex *Holotype* in PDD 30406, Centennial Pk., Upper Piha Valley, Waitakere Ra., Auckland, N.Z., 2.V.1963, J. Dingley (SJH 665).

Known distribution: New Zealand.
For affinities see *C. novae-zelandiae.*

45) **Chalara prolifera** sp. nov. (Figure 23A)
Colonia superficialis, effusa, rufo-brunnea, velutina. Phialophora in fasciculis pusillis, ex tenui stromate orientia, recta vel leniter flexa, multi-septata, atrobrunnea ad basem, insuper pallescentia, paries minute verruculosus et pro ratione crassus, constrictus ad septa; in phialidem terminantia. Phialides lageniformes vel

conicae, pallide brunneae, 29-59 [48] μ long.; venter conicus, 18-28 [24] x 7.5-10 [8.5] μ; collum subcylindraceum vel conicum, 19-36 [30] x 3.5-6.5 [5] μ; transitio ex ventre ad collum abrupta; ratio long. colli et ventris = 1.2:1. Phialoconidia singulatim vel in catenas breves extrusa; cylindracea, apice rotundato, base truncata fimbriam minutam et marginalem ferente; vulgo 1-septata, raro 2-septata, hyalina, pariete laevi; 10-16 [13] x 3-4.5 [3.7] μ; ratio conidii long./lat. = 3.4:1.

Colony superficial, effuse, reddish brown, hairy. Phialophores arising in small clusters from a thin stroma; erect or slightly bent, many-septate, dark brown below, becoming progressively paler above; wall minutely verrucose and relatively thick, constricted at the septa; terminating in a phialide. Phialides lageniform or conical, pale brown, 29-59 [48] μ long; venter conical, 18-28 [24] x 7.5-10 [8.5] μ; collarette subcylindrical or obconical, 19-36 [30] x 3.5-6.5 [5] μ; transition from venter to collarette abrupt; ratio of mean lengths of collarette and venter = 1.2:1. Repeated percurrent proliferation (up to seven times) common. Sympodial proliferation may also occur. Phialoconidia extruded singly or in short chains; cylindrical, rounded at the apex, truncate at the base with a minute but discernible marginal frill, mostly 1-septate, rarely 2-septate, hyaline, smooth-walled; 10-16 [13] x 3-4.5 [3.7] μ; mean conidium length/width ratio = 3.4:1.

Habitat: On twigs of *Thea* sp.

Specimen examined: IMI 111990 [*Holotype*], Mauritius, 11.I.1965, S. Felix.

Known distribution: Mauritius.

C. prolifera is fairly close to *C. selaginellae*, but has a conical venter, subcylindrical or obconical collarette, and usually abrupt transition from venter to collarette.

46) Chalara pteridina Syd. (Figure 14)
in Annls. mycol., 10: 450, 1912.

Colony superficial, effuse, reddish brown, forming irregular patches. Phialophores simple, cylindrical, multi-septate, dark brown; wall smooth and 1 μ thick, slightly constricted at the septa; 46-205 [86] μ long and 4.5-9 [6.3] μ wide at the base; terminating in a phialide. Phialides obclavate to subcylindrical, 31-71 [52] μ long; venter subcylindrical, 13-40 (-46) [27] x 5-9.5 [7.2] μ; collarette cylindrical, 12-44 (-50) [25] x 3-4.5 [3.6] μ; transition from venter to collarette gradual; ratio of mean lengths of collarette and venter = 0.9:1. Percurrent proliferation frequent. Phialoconidia extruded singly or in very short chains; cylindrical, rounded or truncate at the apex, and truncate at the base with a minute but distinct marginal frill; mostly 3-septate (sometimes 0-2-septate), hyaline, smooth-walled; 8-18 [12] x 2-3 [2.5] μ; mean conidium length/width ratio = 5:1.

Habitat: on *Bidens, Pteridium, Senecio, Tanacetum,* dead wood and unidentified herbaceous stems.

Exsiccati examined: S 1144 [*Holotype*], Mycotheca Germanica #1144, on *Pteridium aquilinum*, 7.VIII.1912, P. Sydow (also in IMI 27811, B, BPI, FH); IMI 13191 ex *Chalara brachyspora* folder in K, 'Kryptogamiae exsiccatae editae a Museo Hist. Natur. Vindobonensis 2527, on *Pteridium aquilinum,* Rekawinki, Austria, C. Kiessler.' Also in B, BPI, C and FH.

Specimens examined: 1) IMI 17541, on *Pteridium aquilinum,* Woodbury Common, Devon, U.K., 17.IX. 1963, M. B. Ellis; 2) IMI 61415(b), on *Pteridium aquilinum,* Ranmore Common, Surrey, U.K., XI.1955, C. Booth; 3) DAOM 34484 ex Herb.

IMI 7465, on *Pteridium aquilinum,* Skeldo Plantation, Yorks.,U.K., 15.IX.1946; 4) IMI 74219(a), on *Senecio jacobina,* Stewart Is., Horseshoe Bay, N.Z., II.1954, J. M. Dingley; 5) IMI 89641, on herbaceous stem, Sugar Hill, Swanage, Dorset, U.K., 26.V.1961, K. A. Pirozynski; 6) IMI 95313(b), on *Tanacetum vulgare,* Flatford Mill, Suffolk, U.K., 20.VIII.1962, B.C. Sutton; 7) FH 1624 in folder 11131, on *Bidens* sp., Luxembourg, 3.VIII.1901, J. Feltgen ex Herb. v. Höhnel; 8) PDD 18319 on *Cyperus ustulatus,* Huia, Auckland, N.Z., 12.VI.1953, J.M.D.

Known distribution: Austria, Germany, Luxembourg, New Zealand, U.K.

C. pteridina is close to *C. prolifera* but has larger phialides, subcylindrical venters, and narrower conidia which are 0-3-septate.

47) Chalara pulchra Nag Raj & Hughes (Figure 16)
in N.Z. Jl. Bot., 12: 126, 1974.

Colony discrete, effuse, brown, with cream coloured layer of conidia. Phialophores in loose or compact clusters; simple, composed of sessile, subcylindrical phialides borne directly on the thin basal stroma, or on one-celled stalks 7-9 μ long and 6-9 μ wide; 105-150 [135] μ long, 11-15 [13] μ wide at the median part of the barely differentiated venter, and 9.5-12 [11] μ wide at the apex of the cylindrical collarette; brown to pale brown and smooth-walled. Proliferation not seen. Phialoconidia extruded in readily seceding chains; cylindrical, apex blunt or rounded, base slightly rounded or truncate with distinct marginal frills; predominantly 7-septate (rarely 5-6-septate), hyaline to subhyaline, smooth-walled and without constrictions at the septa; 39-56 [49] x 8.5-11 [10] μ, mean conidium length/width ratio = 5:1.

Habitat: On *Weinmannia racemosa,* unidentified wood.

Specimens examined: 1) DAOM 110020 ex *Holotype* in PDD 30408, Lake Ianthe, Pukekura, Westland, N.Z., 8.IV.1963, S. J. Hughes (568); 2) PDD 21435, on dead wood, Whitianga Rd., Coromandel Pen., Auckland, N.Z., 21.VIII.1963, F. J. Morton.

Known distribution: New Zealand.

This fungus is distinct from all other known species of *Chalara* and appears to be marginal between *Chalara* and *Ascoconidium.*

48) Chalara quercina Henry (Figure 32A)
in Phytopathology, 34: 631, 1944.

Phialophores simple, cylindrical, many-septate, subhyaline, 25-65 [40] μ long, 2.5-4 μ wide at the base; terminating in a phialide. Phialides subcylindrical, hyaline to subhyaline, 25-34 [31] μ long; venter subcylindrical or cylindrical, 3-4.5 [3.6] μ wide; collarette cylindrical, 2.5-4 [2.8] μ wide; transition from venter to collarette almost imperceptible. Proliferation not seen. Phialoconidia extruded singly or in long, easily dispersible chains; cylindrical, unicellular, hyaline; 3.5-8 [5.7] x 2-3.5 [2.5] μ; mean conidium length/width ratio = 2.2:1.

Habitat: On *Quercus.*

Specimen examined: BPI-FP 97476 [*Holotype*].

Chalara quercina resembles *Chalara ungeri,* but has subhyaline phialophores, and smaller, narrower phialides and conidia.

49) Chalara rhynchophiala sp. nov. (Figure 25A)

Colonia superficialis, effusa, lutea vel sulphurea, irregularis et variabilis. Hyphae superficiales subhyalinae vel pallide brunnea, septatae, 2.5-3 μ lat., pariete laevi. Phialophora simplicia, in fascicula arcta aggregata, phialide terminali in brevi unicellulari stipite 2.5-6.5 μ long., portata, vel saepe ad phialidem tantum redacta. Phialides lageniformes vel ampulliformes, brunnea, 37-57 [49] μ long; venter globosus, 9-15 [12] x 7.5-10 [9] μ; collum cylindraceum, 25-42 [36] x 2.5-3.5 [3] μ; transitio ex ventre ad collum abrupta; ratio long. colli et ventris = 3:1. Phialoconidia singulatim vel facile dissilientes catenas extrusa; cylindracea, extremis aliquantum rotundatis vel apice rotundato et base truncata, 1-septata; hyalina, pariete laevi; 12-23 [19] x 2.5-3.5 [2.7] μ; ratio conidii long./lat. = 7:1.

Colony superficial, effuse, yellowish white to sulphur yellow, in irregular and variable patches. Superficial hyphae subhyaline to pale brown, septate at intervals of 9-12 μ; smooth-walled; 2.5-3 μ wide. Phialophores simple, in closely aggregated clusters, as short, unicellular, cylindrical supporting structures 2.5-6.5 μ long, with a terminal phialide; but often reduced to the phialide alone. Phialides lageniform or ampulliform, brown, 37-57 [49] μ long; venter globose, 9-15 [12] x 7.5-10 [9] μ; collarette cylindrical, 25-42 [36] x 2.5-3.5 [3] μ; transition from venter to collarette abrupt; ratio of mean lengths of collarette and venter = 3:1. Proliferation not seen. Phialoconidia extruded singly or in easily dispersible chains; cylindrical with rounded ends, or base truncate and apex rounded; 1-septate, hyaline, smooth-walled; 12-23 [19] x 2.5-3.5 [2.7] μ; mean conidium length/width ratio = 7:1.

Habitat: On *Aesculus* sp., *Dracophyllum* sp.

Specimens examined: 1) IMI 90620 [*Holotype*], on *Aesculus* sp. Dalby Forest, Thornton-le-Dale, Pickering, Yorks., U.K., 22.X.1961, W. G. Bramley; 2) IMI 90621(a), on *Aesculus* sp. Kingthorpe, Pickering, Yorks., U.K., 22.X.1961, W.G.B.; 3) PDD 32867 on *Dracophyllum* sp., Pegleg Cr., Arthur's Pass Nat. Pk., N.Z., 27.IV.1974, B.K. (KNZ 633).

Known distribution: New Zealand, U.K.

C. rhynchophiala resembles *C. fusidioides* but has larger phialides, longer collarettes in proportion to the length of the venters, and larger, 1-septate, conidia.

50) Chalara rostrata sp. nov. (Figure 22B)

Phialophora solitaria vel in fasciculos 3 vel 4 aggregata,ex strato tenui cellularum incrassatarum, atrobrunnearum exorientia; simplicia, cylindracea, multiseptata, brunnea ad atrobrunnea, pariete laevi; 110-195 [145] μ long., 4-6 [4.8] μ lat. ad basem, in phialidem terminantia. Phialides obclavatae, 40-78 [57] μ long.; venter cylindraceus vel ellipsoideus, 18-28 [23] x 6-9.5 [7.5] μ; collum cylindraceum, 26-47 [41] x 3-6 [4] μ; transitio ex ventre ad collum gradatim; ratio long. colli et ventris = 1.7:1. Phialoconidia singulatim vel in catenas extrusa; cylindracea, apice rotundato, base truncata fimbriam marginalem ferente; 1-septata, hyalina, pariete laevi; 9-13 [10] x 2.5-3 [2.6] μ; ratio conidii long./lat. = 4:1.

Phialophores solitary or in clusters of 3-4, arising from a thin layer of thick-walled, dark brown cells on the substratum; simple, cylindrical, multiseptate, brown to dark brown, smooth-walled; 110-195 [145] μ long, 4-6 [4.8] μ wide at the base; terminating in a phialide. Phialides obclavate, 40-78 [57] μ long; venter cylindrical or ellipsoidal, 18-28 [23] x 6-9.5 [7.5] μ; collarette cylindrical, 26-47 [41] x 3-6 [4] μ; transition from venter to collarette gradual; ratio of mean lengths of collarette and venter = 1.7:1. Proliferation rare, percurrent. Phialoconidia extruded singly or in chains; cylindrical, apex rounded, base truncate with a marginal frill, 1-septate, hyaline, smooth-walled; 9-13 [10] x 2.5-3 [2.6] μ; mean conidium length/ width ratio = 4:1.

Habitat: On *Geostachys rupestris.*

Specimen examined: IMI 54897 [*Holotype*] , Cameron Highlands, Malaysia, 6.IX.1953, W. J. Cherewick (Malaya No. 1094).

Known distribution: Malaysia.
   Affinities of *C. rostrata* are noted under *C. acuaria.*

51) Chalara rubi Sacc. & Briard apud Briard (Figure 18B)
in Revue mycol., 8: 24, 1886.
   Phialophores simple, erect or variously bent, cylindrical, up to 7-septate, brown; wall smooth and thick, sometimes moderately constricted at the septa; 53-110 [75] μ long and 4.5-8 [5.8] μ wide at the base; terminating in a phialide. Phialides obclavate or lageniform, 31-78 [48] μ long; venter ellipsoidal or subcylindrical, 49-57 [53] x 7.5-9.5 [9] μ; collarette cylindrical 15-22 [19] x 6-8 [6.5] μ; transition from venter to collarette gradual; ratio of mean lengths of collarette and venter = 0.36:1. Proliferation rare, percurrent. Phialoconidia extruded singly; cylindrical, rounded at the apex, or blunt at both ends, and adorned with frayed fringes of wall material; 1-septate, hyaline; with smooth, relatively thick walls, slightly constricted at the septa; 11-22 [15] x 4-5 [4.7] μ; mean conidium length/width ratio = 3:1.

Habitat: On *Rubus* sp.

Specimen examined: IMI 69882 (slide) ex *Holotype*, on *Rubus*, Forêt de Ramilly-les-Nantes, 14.IX.1885.

Known distribution: France.
   For affinities of *C. rubi* see *C. acuaria.*

52) Chalara scabrida sp. nov. (Figure 19A)
   Colonia effusa, pallide brunnea, caespitosa. Mycelium vegetativum tenuiter aggregatum; hyphae brunneae, scabridae, 2.5-4.5 μ lat. Phialophora vulgo ad phialides redacta,vel raro ex phialidibus et brevi cellula fulcienti composita. Phialides subcylindraceae, raro lageniformes, flavo-brunneae et concolorae, 37-47 [41] μ long.; venter conicus vel raro ellipsoideus, basaliter scabridus, 10-16 [12.8] x 5-9.5 [6.7] μ; collum cylindraceum, raro obconicum, 22-32 [28] x 3-4 [3.3] μ; transitio ex ventre ad collum gradatim; ratio long. colli et ventris = 2.2:1. Phialoconidia in catenas breves extrusa, cylindracea, apice rotundato, base truncata fimbriam marginalem ferente; 1-septata, hyalina, laevia, 15-19 [17] x 2.5-3 [2.7] μ; ratio conidii long./lat. = 6.4:1.
   Colony effuse, pale brown, caespitose. Vegetative mycelium thinly aggregated; hyphae brown, scabrous, 2.5-4 μ wide. Phialophores usually reduced to phialides, or rarely phialides with a short stalk cell. Phialides subcylindrical, rarely lageniform, golden brown and concolorous, 37-47 [41] μ long; venter conic or rarely ellipsoidal, rough-walled, 10-16 [13] x 5-9.5 [6.7] μ; collarette cylindrical, rarely obconical, 22-32 [28] x 3-4 (3.5) μ; transition from venter to collarette gradual; ratio of mean lengths of collarette/venter = 2.2:1. Phialoconidia extruded in short chains; cylindrical, apex rounded, base truncate with distinct marginal frill; 1-septate, hyaline, smooth; 15-19 [17] x 2.5-3 [2.7] μ; mean conidium length/width ratio = 6.4:1.

Habitat: On dead leaves of *Agathis australis.*

Specimen examined: PDD 32857 [*Holotype*], Summit Tr., Little Barrier Is., N.Z., 7.III.1974, B. Kendrick (KNZ 487a).

Known distribution: New Zealand.
    For affinities of *C. scabrida* see *C. curvata.*

53) Chalara selaginellae Farr apud Farr & Horner Jr. (Figure 23B)
in Nova Hedwigia, 15: 269, 1968.
    Basal stroma variable, composed of isodiametric dark brown cells giving rise to phialophores. Phialophores simple, cylindrical, subcylindrical or obclavate; 1-2-(4-)septate, brown; wall smooth and 1 μ thick, unconstricted at the septa; 32-55 [47] μ long, and 3.5-6.5 [5.5] μ wide at the base; terminating in a phialide. Phialides subcylindrical to lageniform, 25-45 [39] μ long; venter subcylindrical or ellipsoidal, 14-17 [16] x 6.5-8.5 [7.3] μ; collarette cylindrical, 25-29 [27] x 4.5-6.5 [5.2] μ; transition from venter to collarette gradual; ratio of mean lengths of collarette and venter = 1.6:1. Proliferation rare, percurrent. Phialoconidia extruded singly; cylindrical, rounded at the apex, truncate at the base and provided with a clearly discernible marginal frill; usually 1-septate, sometimes unicellular; hyaline, subhyaline or pale brown, smooth-walled; 10-18 [13] x 3.5-4 [3.7] μ; mean conidium length/width ratio = 3.5:1.

Habitat: On *Selaginella.*

Specimen examined: BPI,U.S. 1184620 [*Paratype*], on leaves of *Selaginella rupestris,* 4,900 ft., S. ridge of Snowball Mt., Buncombe Co., N.C., U.S.A., 28.VII.1933, Wherry. [*Holotype:* Florida, U.S.A., w/o date, on leaves of *Selaginella arenicola* ssp. *acanthanota,* Harper s.n. (F. 2870420); *Isotype*, BPI.]

Known distribution: U.S.A.
    For affinities of *C. selaginellae* refer to *C. prolifera.*

54) Chalara sessilis sp. nov. (Figure 30E)
    Colonia inconspicua, effusa, straminea, pulveracea. Phialophora ad phialides redacta, raro ex phialidibus singulari, cylindrica, cellula fulcienti usque ad 6 μ long. et 2.5 μ lat., composita. Phialides ampulliformes vel anguste conicae, base lata et plana; 18-23 [21] μ long., pallide brunneae et concolorae; venter plerumque conicus, interdum ellipsoideus, 10-12.5 [11] x 3-4 [3.7] μ; collum cylindraceum, 10-11 [10.3] x 1.5 μ; transitio ex ventre ad collum gradatim, raro abrupta; ratio long. colli et ventris = 0.9:1. Phialoconidia in catenas facile dispergendas extrusa; cylindracea, utrinque truncata, unicellularia, hyalina, laevia, 3.5-6 [4.8] x (1-) 1.5 μ; ratio conidii long./lat. = 3.4:1.
    Colonies inconspicuous, effuse, straw coloured, powdery. Phialophores reduced to phialides, or rarely phialides with a single cylindrical stalk cell up to 6 μ long and 2.5 μ wide. Phialides ampulliform to narrowly conical, with a broad flat base; 18-23 [21] μ long, pale brown and concolorous; venter usually conic, occasionally ellipsoid, 10-12.5 [11] x 3-4 [3.7] μ; collarette cylindrical, 10-11 [10.3] x 1.5 μ; transition from venter to collarette gradual, rarely abrupt; ratio of mean lengths of collarette/venter = 0.9:1. Phialoconidia extruded in easily dispersible chains; cylindrical with truncate ends, unicellular, hyaline, smooth, 3.5-6 [4.8] x (1-) 1.5 μ; mean conidium length/width ratio = 3.4:1.

Habitat: On dead leaves of *Knightia excelsa.*

Specimen examined: PDD 32639 [*Holotype*] , Kauri Glen Pk., Northcote, Auckland, N.Z., 19.II.1974, B. Kendrick (KNZ 372).

Known distribution: New Zealand.

   C. sessilis superficially resembles *C. constricta,*but differs by its ampulliform to narrowly conical and shorter phialides, collarette concolorous with venter, transition unmarked by a constriction in the wall, and cylindrical conidia with truncate ends.

55) Chalara spiralis sp. nov. (Figure 27)

   Colonia superficialis, effusa, candida vel lutea, pubescens, demum lanata et viridi-flava. Phialophora simplicia, solitaria et sparsa vel in fasciculis pusillis, cylindracea, recta vel varie flexa, multi-septata, flavide-brunnea vel pallide brunnea, paries laevis et pro ratione incrassatus, modice constrictus ad septa, 56-135 [88] μ long., 3-5.5 [4.3] μ lat. ad basem, in phialidem terminantia. Phialides subcylindraceae, pallide brunneae, 42-63 [54] μ long.; venter subcylindraceus, 16-28 [24] x 5.5-8 [6.7] μ; collum cylindraceum, 21-37 [31] x 2.5-4.5 [3.4] μ lat.; transitio ex ventre ad collum abrupta, saepe exigua constrictione; ratio long. colli et ventris = 1.27:1. Phialoconidia interdum singulatim extrusa, sed saepius in longissimas et torsivas catenas quae plus minusve hyphis similes apparent; cylindracea, apice rotundato vel obtusa vel utrinque truncata, 1-septata, hyalina, pariete laevi, 17-28 [21] x 2.5-3.5 [2.8] μ; ratio conidii long./lat. = 7.4:1.

   Colony superficial, effuse, dull white or yellowish white, downy; older patches appearing woolly and greenish yellow. Phialophores simple, solitary and scattered or in small clusters; cylindrical, erect, straight or variously bent, multiseptate, yellowish brown or pale brown; wall smooth and relatively thick, moderately constricted at the septa; 56-135 [88] μ long, 3.5-5 [4.3] μ wide at the base; terminating in a phialide. Phialides subcylindrical, pale brown, 42-63 [54] μ long; venter subcylindrical, 16-28 [24] x 5.5-8 [6.7] μ; collarette cylindrical, 21-37 [31] x 2.5-4.5 [3.4] μ; transition from venter to collarette abrupt, often with a mild constriction; ratio of mean lengths of collarette and venter = 1.27:1. Proliferation uncommon, percurrent or sympodial. Phialoconidia sometimes extruded singly, but more often in extremely long, helically coiled chains which appear more or less hyphoid; cylindrical, rounded at apex or blunt or truncate at both ends; 1-septate, hyaline, smooth-walled; 17-28 [21] x 2.5-3.5 [2.8] μ; mean conidium length/width ratio = 7.4:1.

Habitat: On *Fagus* sp.

Specimen examined: IMI 96708 [*Holotype*] , Thornton-le-Dale, Pickering, Yorks., U.K., 15.XI.1962, W. G. Bramley.

Known distribution: U.K.

   For affinities of *C. spiralis* see *C. aurea.*

56) Chalara stipitata sp. nov. (Figure 24B)

   Colonia effusa, brunnea vel atrobrunnea, pubescens. Phialophora solitaria vel laxe aggregata, recta vel varie flexa, 70-120 [98] μ long., 3.5-4 μ lat., cellula basalis sufflata, 6-8 μ lat., multiseptata, septa parte distali tenuiora, infra brunnea vel fusca, supra pallide brunnea vel flava, laevia, in phialidem terminantia. Phialides lageniformes, 30-34 [32] μ long., pallide brunnea vel flava et concolora; venter cylindraceus vel subcylindraceus, 12-16 [14] x 5-6 [5.6] μ; collum cylindraceum, 17-19 [18] x 2.5-3 μ; transitio ex ventre ad collum abrupta; ratio long. colli et ventris = 1.3:1. Phialoconidia in catenis brevibus et fragilibus reperta; cylindracea, apice rotundato, base truncata fimbriam marginalem minutam ferente, 1-septata, hyalina, laevia, 11-15 [13] x 2-2.5 μ; ratio conidii long./lat. = 5.8:1.

Colony effuse, brown to dark brown, hairy. Phialophores solitary or loosely aggregated, erect or variously bent, 70-120 [98] μ long, 3.5-4 μ wide, basal cell swollen to a width of 6-8 μ, multiseptate, septa in the distal part thinner; brown or smoky brown below, pale or yellowish brown above, smooth-walled, terminating in a phialide. Phialides lageniform, 30-34 [32] μ long, pale or yellowish brown and concolorous; venter cylindrical or subcylindrical, 12-16 [14] x 5-6 [5.6] μ; collarette cylindrical, 17-19 [18] x 2.5-3 μ; transition from venter to collarette abrupt; ratio of mean lengths of collarette and venter = 1.3:1. Phialoconidia occurring in short, fragile chains; cylindrical, apex rounded, base truncate with a minute marginal frill; 1-septate, hyaline, smooth-walled; 11-15 [13] x 2-2.5 μ; mean conidium length/width ratio = 5.8:1.

Habitat: On dead leaves of *Agathis australis* and *Podocarpus hallii*.

Specimens examined: All on leaves of *Agathis australis:* 1) PDD 32860, unmarked Tr. off Scenic Dr., Waitakere Ra., Auckland, N.Z., 27.II.1974, B. Kendrick (KNZ 422a); 2) PDD 32859, nr. Te Wairere Str., Little Barrier Is., N.Z., 9.III.1974, B.K. (KNZ 473a); 3) PDD 32638 [*Holotype*] , Atkinson's Pk., Titirangi, Auckland, N.Z., 15.V.1974, B.K. (KNZ 622); 4) on leaf of *Podocarpus hallii*, PDD 32858, Governor's Bush Tr., Mt. Cook Nat. Pk., N.Z., 25.IV.1974, B.K. (KNZ 573).

Known distribution: New Zealand.

For affinities of *C. stipitata*, see *C. aotearoae* and *C. cylindrosperma*.

57) Chalara thielavioides (Peyr.) comb. nov. (Figure 44)
≡ *Chalaropsis thielavioides* Peyr.
in Staz. sper. agr. it., 49: 595, 1916.

Colony superficial, effuse, brown. Vegetative hyphae hyaline or subhyaline, 4.5-5 μ wide. Phialophores hyphoid, erect, simple, cylindrical, up to 7-septate, hyaline, subhyaline or pale brown, walls smooth or minutely verrucose; 45-150 μ long, 4-6.5 μ wide at the base; terminating in a phialide. Phialides subcylindrical to obclavate, 29-65 [47] μ long; venter subcylindrical or slightly ellipsoidal, 14-29 [21] x 3.5-6.5 [5.3] μ; collarette cylindrical, 15-42 [25] x 2.5-4.5 [3.1] μ; transition from venter to collarette gradual; ratio of mean lengths of collarette and venter = 1.2:1. Phialoconidia extruded in long chains; cylindrical with rounded or truncate ends; unicellular, hyaline, subhyaline or pale brown, with walls smooth or minutely verrucose; 6.5-32(-53) [14] x 2.5-6.5 [4] μ; mean conidium length/width ratio = 3.5:1. Chlamydospores solitary and terminal on sympodially branching conidiophores; holoblastic; usually globose, or ellipsoidal, ovoid, or pyriform with truncate or blunt base; unicellular, yellowish brown to brown, smooth-walled; 7.5-17 [14] μ diam., or 9-19 [15] x 7.5-18 [13] μ; germ slit or pore not seen.

Habitat: On *Crotalaria*, *Daucus carota*, *Juglans*, *Lupinus*, *Rosa*, *Ulmus*.

Lectotype: Fig. 1, p. 585, Peyronel in Le Staz. sper. agr. it. 49, 1916.

Specimens examined: 1) PAD 614, Herb. Saccardo, on *Lupinus albus*, Rome, Italy, XII.1916, Peyronel; 2) IMI 14128, on *Daucus*, U.K., 1.1931, comm. Wakefield;

3) IMI 14129, on *Juglans* grafts, East Malling, U.K., 15.VII.1931, comm. H. Wurmold; 4) IMI 33898, on *Ulmus* wood, Ventonvase, Perranzabuloe, Cornwall, U.K., 27.I.1949, R. Rilstone; 5) IMI 47692, on *Daucus,* Norfolk, U.K., J. Ives; 6) IMI 79728, on *Crotalaria juncea,* Segawa Estate, North Borneo, Indonesia, 4.IX.1959, A. Johnston; 7) IMI 105273, on *Daucus carota,* imported from Holland, Richmond, U.K., Waterston; 8) DAOM 44086, on carrots from Texas, U.S.A.; 9) DAOM 84174, on carrots in cold storage, B.C., Canada; 10) DAOM 88210, on carrots, Fort Williams, N.S., Canada, C.O. Gourley #K211.

Known distribution: Canada, Indonesia, Italy, Netherlands, U.K., U.S.A.

Typification: Peyronel did not designate a holotype. The original specimen now bears only the phialoconidia and chlamydospores. Since Peyronel's illustration of the fungus in his Figure 1, p. 585 in Le Staz sper. Agr. it. 49, 1916 shows the characters of the fungus in good detail and forms part of the protologue of the taxon, this figure is chosen as the lectotype of the species.

Cultural characters: On PDA the colonies are restricted (6-10 mm. diam. in 35 days) or spreading in irregular patches with indented margins, caespitose to velutinous, grayish brown at the periphery, black and powdery in the center. On MA, the aerial mycelium is more abundant; the colonies attain a much greater diameter (70 mm. in 35 days), while branching of the hyphae is more profuse, giving rise to dense aggregates of conidiophores.

Sugiyama (1968) described *Chalaropsis thielavioides* var. *ramosissima* Sugiyama as follows: "varietas haec var. *ramosissima* ab var. *thielavioide* differt colonia evoluta, aleuriosporis majoribus, sed praesertim phialophoris multiramosis." He considered this isolate obtained from soil as distinct because of the rapid growth of the colony, large 'aleuriospores' and profuse branching of the phialophores, and gave dimensions as follows: phialoconidia—4-30 (-40) x 2.5-3 μ; chlamydospores—(8-) 10-15 (-20) x 7.5-12 (-15) μ. These dimensions are well within the broad limits of the species. The rate of growth and branching of the conidiophores is largely governed by the cultural conditions and the medium. In our opinion, the variations listed by Sugiyama are largely physiological manifestations and do not warrant the treatment of the isolate as a distinct variety.

Chalara thielavioides is close to *C. ovoidea* and the *Chalara* state of *Ceratocystis fimbriata*, but has predominantly globose and wider chlamydospores, and wider phialoconidia.

58) **Chalara tubifera** sp. nov. (Figure 20C)

Colonia effusa, brunnea vel atrobrunnea, velutina. Hyphae vegetativae pallide brunneae, laeves, usque ad 2.5 μ lat. Phialophora solitaria vel gregaria, recta vel varie flexa, 65-220 μ long., infra 5-6 μ lat., praeter cellulam basalem late conicam, lobatam et usque ad 11 μ lat.; pauci- vel pluri-septata; infra atrobrunnea, supra pallescentia laevia. Phialides urceolatae, raro lageniformes, 49-65 [58] μ long.; venter anguste obconicus, pallide brunneus, laevis, 19-26 [24] x 6-7 [6.6] μ; collum cylindraceum, basaliter fuscum sed supra pallescens, laeve, 28-40 [35] x 2.5-3 [2.8] μ; transitio ex ventre ad collum abrupta; ratio long. colli et ventris = 1.4:1. Phialoconidia singulatim vel in catenas breves extrusa; cylindracea, apice rotundato, base truncata, fimbriam minutam marginalem ferente; 1-septata, hyalina, laevia; 11-19 [16] x 2-2.5 [2.3] μ; ratio conidii long./lat. = 6.7:1.

Colony effuse, brown to dark brown, hairy. Vegetative hyphae pale brown, smooth, up to 2.5 μ wide. Phialophores solitary or gregarious, erect or variously bent, 65-220 μ long, 5-6 μ wide below, except for the basal cell which is broadly conic, lobed and up to 11 μ wide; few- to many-septate; dark brown below, becoming lighter above, smooth. Phialides urceolate, rarely lageniform, 49-65 [58] μ long; venter narrowly obconic, pale brown, smooth, 19-26 [24] x 6-7 [6.6] μ; collarette cylindrical, brown in the basal part but paler at its apex, smooth, 28-40 [35] x 2.5-3 [2.8] μ; transition from venter to collarette abrupt; ratio of mean lengths of collarette/venter = 1.4:1. Phialoconidia extruded singly or in short chains; cylindrical, apex rounded, base truncate with a minute marginal frill; 1-septate, hyaline, smooth; 11-19 [16] x 2-2.5 [2.3] μ; mean conidium length/width ratio = 6.7:1.

Habitat: On dead leaves of *Knightia excelsa.*

Specimen examined: PDD 32871 [*Holotype*] Hongi's Tr. Scenic Reserve, N.Z., 12.I.1974, B. Kendrick (KNZ 136).

Known distribution: New Zealand.
   C. tubifera resembles *C. urceolata* but has smaller phialides, a shorter, narrower and basally dark collarette, and narrower conidia bearing basal frills.

59) Chalara ungeri Sacc. (Figure 32B)
in Sylloge Fung., 4: 336, 1886.
   Phialophores simple, cylindrical, erect, straight or variously bent, multiseptate, pale to dark brown; 100-900 μ long, 4.5-9 μ wide at the base; terminating in a phialide. Phialides subcylindrical, 50-61 [58] μ long; venter subcylindrical, 4.5-7.5 [6.2] μ wide, collarette more or less cylindrical, 4.5-6.5 [6] μ wide; transition from venter to collarette almost imperceptible. Phialoconidia extruded in short chains or singly; cylindrical or doliiform with blunt ends; unicellular, hyaline, smooth-walled; 5.5-11 [7.8] x 3.5-4.5 [4.3] μ; mean conidium length/width ratio = 1.8:1.

Habitat: On wood of *Pinus, Pseudotsuga menziesii.*

Specimens examined: 1) IMI 20164 [*Neotype*], on 'blued' pine board, Scotland, M. Wilson (authentic for the name *Ceratocystis coerulescens* [Münch] Bakshi); 2) IMI 32291, isol. ex *Pinus* log, Watford, Herts., U.K., 25.XI.1948, W. O. Harper; 3) IMI 47571, isol. ex *Pseudotsuga menziesii*, B.C., Canada, D. Wells; 4) IMI 56851, J. Hunt #25; 5) IMI 56862, on *Pseudotsuga menziesii*, Bellingham, Wash., U.S.A., J. Hunt #144; 6) IMI 56865, on conifer wood, Oregon, U.S.A., J. Hunt #150.

Typification: Münch (1907) demonstrated the conidial state of *Ceratocystis coerulescens* to be *Chalara ungeri* Sacc., which was based on a collection determined by Unger (1847) as *Graphium penicillioides* Cda. f. *ungeri.* The whereabouts of Unger's collection of the fungus are not known: all attempts to trace it in Austrian Herbaria have been futile. There is little reason to expect that the specimen still exists. IMI 20164 is here designated as neotype of *Chalara ungeri.*
   Davidson (1944) described *Endoconidiophora virescens* Davids. from hardwood lumber in U.S.A. Hunt (1956) considered that there was little morphological difference between *E. virescens* and *Ceratocystis coerulescens*, and treated the two as synonymous. We have examined IMI 25320 ex USDA 94255 ex R. W. Davidson, authentic for the name *E. virescens.* The phialophores in this fungus are simple, cylindrical, septate, yellowish brown, 90-720 μ long, 4-7.5 μ wide at the base; terminating in a phialide. The phialides are subcylindrical to rather obclavate, 52-85 [71] μ long, venter subcylindrical or conical, 19-45 [28] x 2.5-4 [3] μ; collarette subcylindrical, 25-56 [43] x 2.5-4 μ; transition from venter to collarette gradual or occasionally abrupt; ratio of mean lengths of collarette and venter = 1.5:1. Phialoconidia extruded singly or in chains; cylindrical with rounded or truncate ends,

unicellular, hyaline, smooth-walled, 5-19 (-29) [10] x 2-4 [2.7] μ; mean conidium length/width ratio = 3.7:1.

These features indicate that *C. coerulescens* and *C. virescens* manifest morphological differences in their conidial states, and further studies of a large number of collections may demonstrate the need for maintaining the taxa as distinct.

For affinities of *Chalara ungeri,* see *Chalara quercina.*

60) **Chalara unicolor** Hughes apud Nag Raj & Hughes (Figure 13A)
in N.Z. Jl. Bot., 12: 121, 1974.

Colony superficial, effuse, brown, hairy. Phialophores solitary and scattered, erect, subcylindrical, 1-2-(5-)septate in the basal part; uniformly brown; wall smooth and without constrictions at the septa; 115-155 [135] μ long, 8.3-11 μ wide at the base; terminating in a phialide. Phialides subcylindrical, 66-140 [100] μ long; venter subcylindrical, 30-63 [47] x 9.5-16 [12] μ; collarette cylindrical, 40-77 [58] x 5.5-9 [7.3] μ; transition from venter to collarette gradual; ratio of mean lengths of collarette and venter = 1.25:1. Phialoconidia extruded in chains; cylindrical, with a bluntly rounded apex and a truncate base bearing a minute but distinct marginal frill; 3-septate, hyaline, smooth-walled; 18-42 (-54) [30] x 5-8 [6.3] μ; mean conidium length/width ratio = 4.7:1.

Habitat: On *Leptospermum scoparium* and undetermined wood.

Specimens examined: 1) DAOM 51641, on undetermined wood, Camp Oswegatchie, nr. Belfort, N.Y., U.S.A., 7.X.1956, S. J. Hughes; 2) DAOM 110019 ex *Holotype* in PDD 30409, on *Leptospermum scoparium,*Kitikiti Stream, Auckland Prov., N.Z., 31.I.1963, S.J.H. (226).

Known distribution: New Zealand, U.S.A.

C. unicolor superficially resembles *C. insignis,* but has uniformly brown phialophores, usually shorter collarettes, and 3-septate rather than 7-septate conidia.

61) **Chalara urceolata** Nag Raj & Kendrick (Figure 26A)
in Nag Raj & Hughes, N.Z. Jl. Bot., 12: 120, 1974.

Colony superficial, effuse, brown, velutinous. Phialophores arising from a thin, superficial, basal stroma; scattered or in small clusters; simple, erect or variously bent, cylindrical, many-septate, brown to dark brown or reddish brown below, paler above; 78-230 [135] μ long and 4.5-8 [6.5] μ wide at the slightly inflated base; terminating in a phialide. Phialides urceolate, pale brown, 28-105 [69] μ long, with smooth or sometimes verrucose walls; venter obconic, 18-37 [26] μ long, narrow at the base and 8-13 [11] μ wide at its apex; collarette 21-72 [52] μ long, 3-6 [5] μ wide; transition from venter to collarette abrupt; ratio of mean lengths of collarette and venter= 2:1. Proliferation rare, percurrent or sympodial. Phialoconidia extruded singly or in fairly persistent chains; cylindrical, both ends rounded or base truncate; 1-septate, hyaline, thin- and smooth-walled; 10-18 [14] x 2.5-4 [3.4] μ; mean conidium length/width ratio = 4.2:1.

Habitat: On *Rhopalostylis sapida, Rumex* sp., unidentified herbaceous stem, and an unidentified umbellifer.

Specimens examined: 1) IMI 5901, on undet. umbellifer, Ashstead, U.K., 23.VI.1946; 2) IMI 31315, on herbaceous stem, Clandeboye, Ireland, 14.IX.1948, E. W. Mason; 3) IMI 102891 [*Holotype*] , on *Rumex* sp., Pickering, Yorks., U.K., 7.VII.1963, W. G. Bramley; 4) DAOM 110039, on *Rhopalostylis sapida,* Cossy's

Creek Dam, Hunua, Auckland Prov., N.Z., 12.II.1963, J. M. Dingley (SJH 313) (in PDD 20529).

Known distribution: Ireland, New Zealand, U.K.
    For affinities of *C. urceolata,* see *C. tubifera.*

62) Chalara state of Ceratocystis adiposa (Butl.) C. Moreau (Figure 37)
in Revue Mycol., 17: 22, 1952.

    Phialophores simple, cylindrical or subcylindrical; few- to many-septate, hyaline, subhyaline or pale brown; wall mostly coarsely verrucose, less than 1 μ thick and slightly constricted at the septa; 33-96 μ long, 4-6.5 μ wide at the base; terminating in a phialide. Phialides subcylindrical or lageniform, 20-40 [30] μ long, often verrucose; venter subcylindrical or ellipsoidal, 12-25 [16] x 4.5-6.5 [5.5] μ; collarette cylindrical, 9-20 [15] x 3-4.5 [3.7] μ; transition from venter to collarette gradual or sometimes abrupt; ratio of mean lengths of collarette and venter = 0.9:1. Phialoconidia mostly in long chains; cylindrical to doliiform with truncate or rounded ends, ovoid, or pyriform with a truncate base; unicellular, hyaline, subhyaline or pale brown, wall smooth or verrucose; 8.5-18 [12] x 3-4.5 (-7) [3.7] μ; mean conidium length/width ratio = 3.3:1. Chlamydospores predominantly enteroblastic-phialidic and sometimes holoblastic; globose, ovoid, or oblong, unicellular, brown to reddish brown, 16-25 [19] μ in diameter or 14-25 [18] x 11-23 [16] μ; with verrucose or fimbriate walls up to 1.5 μ thick; germ slits obscure.

Habitat: On a wide variety of substrates.

Specimens examined: 1) IMI 21355 [*Holotype*], isol. ex sugarcane, Pusa, India, recd. at IMI 15.I.1927, McRae; 2) IMI 21285, isol. ex book back, U.K., recd. at IMI XII.1947, F. Armitage; 3) IMI 56859 J. Hunt #73; 4) IMI 99465, ex wood chips, pulp mill, Mersey, N.S., Canada, 1962, P. Russell.

Known distribution: Widespread.
    The morphology of the chlamydospores and the often verrucose phialides and phialoconidia are distinctive features of this fungus.

63) Chalara state of Ceratocystis autographa Bakshi (Figure 28C)
in Ann. Bot. n.s. Lond., 15: 58, 1951.

    Aerial mycelium funiculose, hyaline and smooth but later fuliginous and rough; hyphae up to 2.5 μ wide. Phialophores arising from aerial mycelium, up to 42 μ long, but usually reduced to sessile phialides, with occasional linear proliferations. Phialides lageniform, 14-20 [16.5] μ long; venter conical, less often ellipsoid, 8-12 [9.5] x 2.5-3.5 [3.1] μ; collarette narrowly obconical, rarely subcylindrical, 6-8 [7] x 1.5-2 μ, with slightly darker and thicker wall than that of the venter; transition from venter to collarette abrupt and usually marked by a constriction; ratio of mean lengths of collarette and venter = 0.7:1. Phialoconidia catenate; short-clavate, apex rounded, base truncate, unicellular, hyaline, 3.5-4 [3.8] x 1-1.5 μ, mean conidium length/width ratio = 3:1.

Habitat: In galleries of *Dryocoetes autographa* and *Hylurgops palliatus* infesting *Larix leptolepis.*

Holotype: IMI 20162.

Cultures examined: Slides ex culture derived from *Holotype.*

Known distribution: U.K.

Bakshi (*op. cit.*) reported the phialoconidia as barrel-shaped and 3.5-6.8 x 1-2 μ, and also noted a second type of ovoid or globose conidia borne on short conidiophores. The young conidia are, in fact, short-clavate with truncate bases. Although we did see the second type of conidia in the slides, we were unable to ascertain the nature of the conidium ontogeny involved. The conidia themselves were hyaline and rather thick-walled. Bakshi did not report the slight but highly characteristic constriction at the base of the collarette.

The *Chalara* state of *C. autographa* is close to *C. constricta,* but has sessile phialides, conical venter, narrowly obconic collarette thickened at its base, and shorter, narrower conidia.

64) Chalara state of **Ceratocystis coerulescens** (Münch) Bakshi: see *Chalara ungeri.*

65) Chalara state of **Ceratocystis fagacearum** (Bretz) Hunt: see *Chalara quercina.*

66) Chalara state of **Ceratocystis fimbriata** Ell. & Halst. (Figure 45)
in Bull. New Jers. agric. Exp. Stn., 76: 14, 1890; J. Mycol., 7: 1, 1891.

Phialophores simple, cylindrical, 1-5-septate, subhyaline to pale brown at the base, becoming progressively lighter toward the distal end; 45-110 μ long, 3-6.5 μ wide at the base; with smooth walls less than 0.5 μ thick, occasionally constricted at the septa; terminating in a phialide. Phialides subcylindrical to lageniform, 33-63 [48] μ long; venter cylindrical, subcylindrical or ellipsoidal, 14-33 [22] x 5-7.5 [6.4] μ; collarette cylindrical; 16-37 [26] x 2.5-4.5 [3.8] μ; transition from venter to collarette gradual; ratio of mean lengths of collarette and venter = 1.18:1. Phialoconidia usually extruded in long chains; cylindrical with blunt or truncate ends, unicellular, hyaline, with smooth walls less than 0.5 μ thick; subhyaline or pale brown; 7-23 (-52) [17] x 2.5-4 (-7) [3.3] μ; mean conidium length/width ratio = 5:1. Chlamydospores usually ellipsoidal or subglobose, doliiform, or pyriform with a truncate base; unicellular, brown to dark brown; 9-11 [10] μ diam., or 10-20 [15] x 7.5-13 [10] μ; with smooth walls up to 1.5 μ thick; germ slit not observed.

Habitat: On *Hevea, Pimenta, Theobroma,* and other angiosperms.

Specimens examined: 1) IMI 25319, authentic for the name *Endoconidiophora fimbriata,* presumed isol. ex brown-stained *Fagus* board; 2) IMI 28618, as *Rostrella coffeae* Zimm. ex coffee stem, Colombia, R. E. Pontis; 3) IMI 86754 ex *Pimenta officinalis,* Hope, Jamaica, West Indies, 7.III.1961, R. I. Leather; 4) IMI 108697, isol. ex *Hevea brasiliensis,* Malaysia, 8.IX.1964, comm. K. P. John; 5) IMI 123638, ex *Theobroma cacao*, River Estate, Trinidad, 1966, E. F. Itou in BPI.

Known distribution: Colombia, Malaysia, West Indies.

Affinities of this species are mentioned under C. *thielavioides.*

67) Chalara state of **Ceratocystis moniliformis** (Hedgc.) C. Moreau (Figure 43A)
in Revue Mycol., 17: 22, 1952.

Phialophores cylindrical, 1-2-septate or often reduced to phialides; hyaline with smooth or occasionally verrucose walls; 21-39 μ long, 2.5-4 μ wide at the base. Phialides subcylindrical to lageniform, 20-36 [26] μ long; venter subcylindrical or ellipsoidal, 11-20 [16] x 2.5-4.5 [3.9] μ; collarette cylindrical, 7.5-14 [10] x 2-3 [2.4] μ; transition from venter to collarette gradual; ratio of mean lengths of collarette and venter = 0.65:1. Phialoconidia extruded in chains; cylindrical with blunt or truncate ends; unicellular, hyaline, smooth-walled; 6-22 [12] x 1-4 [2.4] μ; mean conidium length/width ratio = 5:1. Chlamydospores occurring in long chains, enter-oblastic-phialidic, doliiform, pale brown, smooth-walled; 6-9 x 4-6 μ.

Habitat: wood.

Holotype in BPI.

Specimens examined: 1) IMI 20163 (lectotype of *C. wilsonii*), on *Quercus* sp., ex orig. isol. by M. Wilson, 17.VI.1947; 2) IMI 56860, J. Hunt #75 = B34 of K. Aoshima; 3) IMI 125922, isol. ex *Saccharum* rhizosphere, Mandeville, Jamaica, West Indies, comm. 13.II.1969, B. M. Hogg.

The *Chalara* state of *Ceratocystis moniliformis* has the smallest phialides and phialoconidia of any *Chalara* known to produce chlamydospores and to have *Ceratocystis* as ascigerous state.

Known distribution: widespread.

68) Chalara state of **Ceratocystis paradoxa** (Dada) C. Moreau:
see *Chalara paradoxa*.

69) Chalara state of **Ceratocystis radicicola** (Bliss) C. Moreau (Figure 38)
in Revue Mycol., 17: 22, 1952.

Phialophores simple, cylindrical, septate, hyaline to subhyaline or pale brown, minutely verrucose; 3-6.5 μ wide at the base; terminating in a phialide. Phialides subcylindrical to lageniform, 48-95 [67] μ long; venter subcylindrical, 33-66 [45] x 4.5-6.5 [5] μ; collarette cylindrical, 18-27 [21] x 2-3 [2.8] μ; transition from venter to collarette gradual or often imperceptible; ratio of mean lengths of collarette and venter = 0.47:1. Phialoconidia extruded singly or in chains; cylindrical with blunt or rounded ends; unicellular, hyaline or subhyaline; 7.5-20 [12] x 2-3 [2.5] μ; mean conidium length/width ratio = 4.8:1. Chlamydospores occurring terminally and solitarily on short, septate, hyaline, sympodially branching conidiophores; predominantly holoblastic; ovoid or pyriform with a truncate base, unicellular, brown to dark brown with verrucose or striate walls; 14-22 [17] x 11-15 [13] μ; germ slit vertical.

Habitat: On date palm *(Phoenix)*.

Holotype in BPI.

Specimen examined: IMI 36479, cultured on sterile *Pinus* board at 28 °C in darkness, 27.VI.1968 - 19.IX.1968.

Hennebert (1967) described *Chalaropsis punctulata* Henneb. and noted that it differed from the *Chalaropsis* state of *Ceratocystis radicicola* in the dimensions of its 'aleuriospores' (chlamydospores). The dimensions of phialoconidia and chlamydospores were given as 7-12 x 3-6 μ and 12-20 x 8-13 μ, respectively. These dimensions appear to fall within the broad limits of the *Chalara* state of *Ceratocystis radicicola,* and *Chalaropsis punctulata* may be considered a synonym of this state.

The *Chalara* state of *Ceratocystis radicicola* resembles *Chalara thielavioides,* but has larger phialides, and verrucose or striate chlamydospores.

70) Chalara state of **Cryptendoxyla hypophloia** Malloch et Cain. (Figure 30A)
in Can. J. Bot., 48: 1816, 1970.

Phialophores arising from hyaline hyphae; mostly simple, but rarely branched, 1-2-septate and 30-40 μ long, often reduced to sessile phialides separated from the supporting mycelium by a basal septum; hyaline, smooth-walled. Phialides subcylindrical, erect or often bent, 13-28 [21] μ long; venter subcylindrical to cylindrical, 7.5-14 [11] x 2-2.5 μ; collarette cylindrical, 5.5-11 [9] x 1.5-2 μ; transition from venter to collarette gradual; ratio of mean lengths of collarette/venter = 0.8:1. Phialoconidia catenulate, cylindrical, both ends truncate or base alone truncate; unicellular, hyaline, smooth-walled; 3.5-11 [6] x 1.5-2.5 [1.8] μ; mean conidium length/width ratio = 3.3:1.

Habitat: On wood of *Acer saccharum, Betula lutea.*

Holotype of *Cryptendoxyla hypophloia* in TRTC 45320.

Cultures examined: (grown on sterile *Acer saccharum* bark 5.II.1975 - 12.II.1975): 1) ex DAOM 147382a, under bark of *Acer saccharum,* nr. Star Lake, St. Lawrence Co., N.Y., U.S.A., 8.IX.1974, S. J. Hughes; 2) ex DAOM 147685, under bark of dead, standing *Betula lutea,* Moorside, Gatineau Pk., Gatineau Co., P.Q., Canada, 15.IX.1974, D. Malloch (both authentic for the name).

Known distribution: Canada, U.S.A.

The *Chalara* state of *Cryptendoxyla hypophloia* resembles *Chalara microspora* and the *Chalara* state of *Ceratocystis moniliformis.* It differs from the latter in its constant association with the perfect state, its lack of chlamydospores, and in its smaller phialides and conidia. It differs from *Chalara microspora* in its smaller, hyaline phialophores.

Fusichalara Hughes & Nag Raj
in N.Z. Jl. Bot., 11: 662, 1973.

Phialophores simple, scattered or aggregated into compact fascicles, cylindrical, septate, brown to dark brown, terminating in a phialide. Phialides subcylindrical, obclavate or ampulliform, composed of a venter and a long collarette; transition from venter to collarette gradual or abrupt and marked by a convex thickening of the inner wall of the phialide. Conidia enteroblastic-phialidic, of two kinds: first-formed conidia long-cylindrical, multiseptate, apical cell rounded at the apex, basal cell obconic or cylindrical with or without marginal frills, hyaline to pale brown; subsequent conidia fusiform, straight or slightly sigmoid, septate, hyaline or pale brown to brown with paler end cells.

Type species: Fusichalara dimorphospora Hughes & Nag Raj.

Fusichalara shows considerable affinity with *Chalara* but is distinguished from it by (a) a convex thickening of the inner wall of the phialide at the point of transition from venter to collarette; and (b) the two kinds of phialoconidia, the first-formed being cylindrical, twice as long as, and with about twice as many septa as the subsequent fusiform or slightly sigmoid conidia.

Key to Species
1) Conidia* predominantly
   3-septate . . . .**Fusichalara dingleyae** (2)
1) Conida predominantly 7-septate 2

    2) Conidia mostly 5-6.5 μ wide and 24-47 μ
      long . . . .**Fusichalara novae-zelandiae** (3)
    2) Conidia mostly 7.5-10 μ wide and 30-40 μ
      long . . . . **Fusichalara dimorphospora** (1)

1) Fusichalara dimorphospora Hughes & Nag Raj (Figure 46C)
in N.Z. Jl. Bot., 11: 663, 1973.

Colony superficial, effuse, black, hairy. Phialophores solitary and scattered or in aggregates of 2-4; simple, erect, more or less cylindrical, 3-many-septate, brown or dark brown, with smooth walls 1-1.5 μ thick; 180-310 [260] μ long, usually inflated and up to 15 μ wide at the base, 8-11 μ wide above, terminating in a phialide. Phialides subcylindrical to obclavate, brown to dark brown, with a diffuse darker brown zone around the base of the collarette and also below the paler, torn and frilled apex; 115-180 [150] μ long, composed of a slightly inflated venter 37-53 [44] x 10-14 [12] μ, and a subcylindrical collarette 70-130 [105] μ long and 9-15 [12] μ wide at the apex; transition from venter to collarette gradual; ratio of mean lengths of collarette and venter = 2.4:1. Conidia enteroblastic-phialidic, of two kinds. The first conidium is cylindrical, somewhat rounded at the apex and obconical at the base, 11-17-septate, with septa mostly transverse but one or a few occasionally oblique or longitudinal; pale brown to brown except for the hyaline or subhyaline basal cell; wall smooth and not constricted at the septa; 85-125 [100] x 6-9 [7.8] μ; mean conidium length/width ratio = 13:1. Subsequent conidia are fusiform or slightly sigmoid, with the basal cell more tapered than the bluntly rounded apical cell; predominantly 7-septate but occasionally 3-, 5-, or more rarely 8-septate, septa thick and dark, end cells hyaline to subhyaline, penultimate cells occasionally paler than the uniformly brown median cells; 35-75 [61] x 7.5-10 [8.6] μ; mean conidium length/width ratio = 7:1.

*   Conidia refers to those produced after the first-formed.

Habitat: On bark of dead *Weinmannia racemosa.*

Specimen examined: PDD 30402 [*Holotype*] , L. Ianthe, Pukekura, Westland, N.Z., 8.IV.1963, S. J. Hughes (560i). Also in DAOM 96020(i).

Known distribution: New Zealand.

Affinities of this species are considered under *F. novae-zelandiae.*

A) *Fusichalara dingleyae.* Phialophores, cylindrical and fusiform conidia ex DAOM 82810.
B) *Fusichalara novae-zelandiae.* Phialophores, cylindrical and fusiform conidia ex DAOM 110040.
C) *Fusichalara dimorphospora.* Phialophores, cylindrical and fusiform conidia ex DAOM 96020i.

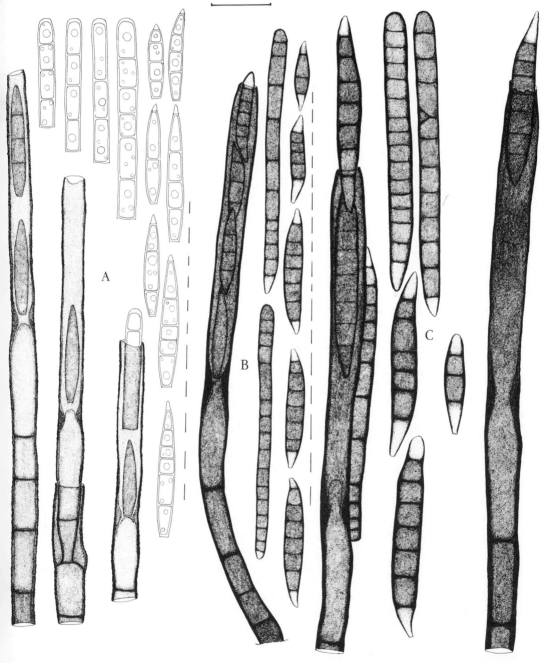

2) Fusichalara dingleyae Hughes & Nag Raj (Figure 46A)
in N.Z. Jl. Bot., 11: 665, 1973.

Colony sparse, effuse, brown, hairy. Phialophores arising from a thin stroma in the substrate, usually in compact fascicles of 15 or more, rarely solitary; cylindrical, 5-7-septate, brown to dark brown in the basal part, becoming progressively paler toward the apex, wall coarsely roughened and 2-2.5 μ thick; 140-305 μ long, 6-10 μ wide at the base, terminating in a phialide. Phialides subcylindrical to ampulliform, yellow to yellowish brown, 90-165 [130] μ long, composed of a venter (21-) 25-47 [38] x (7.5-) 9.5-13 [11] μ, and a cylindrical collarette 65-110 [90] μ long and (6.5-) 7.5-9.5 [8.5] μ wide; transition from venter to collarette gradual or abrupt; ratio of mean lengths of collarette and venter = 2.4:1. Conidia enteroblastic-phialidic, of two kinds. The first conidium long-cylindrical, rounded at the apex and truncate at the base with a clearly discernible marginal frill, hyaline, 0-16-(mostly 7-)septate, with smooth walls not constricted at the septa; 35-95 [65] x 4.5-6.5 [5.6] μ; mean conidium length/width ratio = 12:1. Subsequent conidia occurring in readily seceding chains; fusiform, conical at the apex, truncate at the base with a conspicuous marginal frill, hyaline to subhyaline, 1-5-(mostly 3-)septate, wall smooth and not constricted at the septa; 25-60 [42] x 4-6 [5] μ; mean conidium length/width ratio = 8:1.

Habitat: On *Melicytus ramiflorus, Olearia rami* and undetermined rotten wood.

Specimens examined: 1) PDD 30403, on *Melicytus ramiflorus,* Anawhata Rd., Waitakere Ranges, Auckland, N.Z., 3.X.1963, J. M. Dingley (SJH 1460b) (also in DAOM 93954[b]); 2) DAOM 93595(c), on *Olearia rami,* Walker Bush, Henderson Valley, Waitakere Ra., Auckland, N.Z., 24.X.1963, S. J. Hughes (1564c): all on unidentified wood: 3) DAOM 93569(b), Whitianga Rd., Coromandel Pen., N.Z., 21.VIII.1963, S.J.H.; 4) PDD 21599 [*Holotype*] Home Track, Upper Piha Valley, Waitakere Ra., Auckland, N.Z., 9.X.1963, S.J.H. (1499a) (also in DAOM 93957[a]); 5) DAOM 109353(b), Kauri Knoll Track, Waitakere Ra., Auckland, N.Z., 6.II.1963, S.J.H. (276b); 6) DAOM 109385(b), Walker Bush, Henderson Valley, Waitakere Ra., Auckland, N.Z., 24.X.1963, S.J.H. (1567b); 7)DAOM 110202, Lower Poerua R., Hari Hari, Westland, N.Z., 5.IV.1963, J. M. Dingley; 8) DAOM 110203(b), Titirangi, Auckland, N.Z., 22.I.1963, S.J.H. (196b); 9) DAOM 82810, nr. Onoville, N.Y., U.S.A., 10.VI.1961, S.J.H.

Known distribution: New Zealand, U.S.A.

3) Fusichalara novae-zelandiae Hughes & Nag Raj (Figure 46B)
in N.Z. Jl. Bot., 11: 670, 1973.

Colony superficial, effuse, black, hairy. Phialophores solitary, scattered or densely crowded, simple, erect, cylindrical, 2-5-septate, brown to dark brown, with smooth walls 1-1.5 μ thick; 150-230 μ long, inflated at the base to a width of 16 μ, soon narrowing above to 6.5-9 μ wide, and terminating in a phialide. Phialides subcylindrical to obclavate, brown, 95-155 [130] μ long, composed of a slightly swollen venter 29-40 [35] x 7.5-10 [9.4] μ, and a cylindrical collarette 60-120 [93] μ long and 7-11 [8.3] μ wide; transition from venter to collarette gradual; ratio of mean lengths of collarette and venter = 2.6:1. Conidia enteroblastic-phialidic, of two kinds. The first-formed conidia long-cylindrical with a rounded apex and an obconical base, 8-12-septate, rarely 18-septate, subhyaline to pale brown except for the hyaline or subhyaline basal cell, wall smooth and not constricted at the septa, 65-105 [88] x 4.5-6 [5.3] μ; mean conidium length/width ratio = 16:1. Subsequent conidia fusiform, straight or slightly sigmoid, 3-7-(mostly 7-)septate, end cells hyaline, median cells pale brown; smooth-walled; 24-47 [35] x 5-6.5 [5.4] μ; mean conidium length/width ratio = 6.4:1.

Habitat: On rotten wood of *Leptospermum scoparium.*

Specimen examined: PDD 30404 [*Holotype*] (DAOM 110040), Cornwallis, Auckland, N.Z., 3.I.1963, S. J. Hughes (50).

Known distribution: New Zealand.

    F. novae-zelandiae is similar to *F. dimorphospora,* but distinguishable by (a) its smaller phialophores and conidia, and (b) its subhyaline to pale brown first-formed conidia.

Chaetochalara Sutton & Pirozynski
in Trans. Br. mycol. Soc., 48: 350, 1965.

Colonies foliicolous, effuse, dark brown to black, hairy to setose. Mycelium partly superficial, partly immersed in the substratum. Internal mycelium of hyaline hyphae, often aggregated in the substomatal cavities and emerging through stomata to give rise to tufts of setae and conidiophores, then spreading to form a superficial network of subhyaline to brown hyphae often bearing phialophores, or both setae and phialophores. Setae simple, erect, brown, non-septate or septate, thick-walled and often pointed at the apex. Phialophores mostly reduced to phialides borne directly on the superficial hyphae; but sometimes simple, or less frequently irregularly branched, erect, cylindrical, few- to many-septate, light to dark brown supporting structures are present. Phialides colourless or brown, subcylindrical, cylindrical or ampulliform. Conidia enteroblastic-phialidic, cylindrical, unicellular or septate, hyaline.

Type species: Chaetochalara bulbosa Sutton & Pirozynski.

The genus has a close affinity with *Chalara* and *Sporoschisma*. It differs from *Chalara* only in the presence of sterile setae, and from *Sporoschisma* in its hyaline, unicellular to few-septate conidia and its proportionately larger setae which lack a mucilaginous cap.

Key to Species
1) Conidia unicellular 2
1) Conidia septate 3

   2) Phialides narrowly ampulliform, venter widest at its base, conidia 5-9 x 1.5-2 μ .... **Chaetochalara africana** (1)
   2) Phialides ampulliform, venter bulbous, widest in its median part, conidia 6.5-10 x 1-2 μ .... **Chaetochalara bulbosa** (3)

3) Phialide walls smooth 4
3) Collarette wall sparsely asperate, phialide terminal on a septate, cylindrical stalk, conidia 10-27.5 x 3-4 μ ....**Chaetochalara aspera** (2)

   4) Phialides ampulliform, venter bulbous, conidia 11.5-15.5 x 1.5-2.5 μ ....**Chaetochalara setosa** (6)
   4) Phialides subcylindrical or obclavate 5

5) Phialides sessile or borne on a short basal cell, venter 4.5-9 μ wide, conidia 10-18 x 4-5 μ ....**Chaetochalara cladii** (4)
5) Phialides arising as branches of short phialophores, subcylindrical, venter 3-4.5 μ wide, conidia 9-14 x 3-4.5 μ.....................**Chaetochalara ramosa** (5)

1) Chaetochalara africana Sutton & Pirozynski (Figure 47A)
in Trans. Br. mycol. Soc., 48: 352, 1965.

Colonies amphigenous, effuse, brownish black, hairy, scattered in irregular patches. Fertile hyphae superficial, subhyaline to pale brown, septate, sparingly branched, bearing setae and phialides. Setae simple, erect, 70-150 [122] μ long, expanded below to a width of 5-8 [6] μ, gradually tapering toward the apex, which is 1-2 [1.4] μ wide; non-septate, dark brown tending to become paler toward the distal end, and smooth-walled with occasional intercalary nodulose swellings. Phialides arising directly from the fertile hyphae, often aggregated around a seta, narrowly ampulliform, yellowish brown to brown, 21-37 [30] μ long, maximum

A) *Chaetochalara africana.* Setae, phialophores and conidia ex IMI 94904f.
B) *Chaetochalara bulbosa.* Setae, phialophores and conidia ex IMI 89645a.

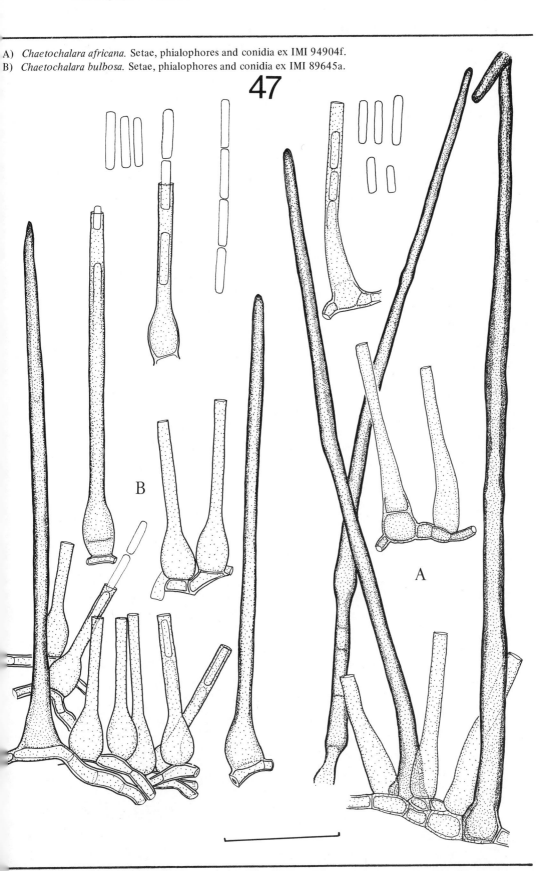

width 2.5-6 [4.8] μ at the base, with a long cylindrical collarette, 2-2.5 [2.3] μ wide, smooth-walled; transition from venter to collarette gradual. Phialoconidia cylindrical with truncate or obtuse ends, non-septate, hyaline, smooth-walled, 5-9 [7.2] x 1.5-2 [1.7] μ; mean conidium length/width ratio = 4:1.

48) *Chaetochalara aspera.* Setae, phialophores and conidia ex DAOM 127892.

48

Habitat: On rotting leaves of *Beilschmiedia tawa* and *Brachystegia spiciformis.*

Specimens examined: 1) IMI 94904(f) [*Holotype*], on *Brachystegia spiciformis,* Mt. Makulu Res. Sta., Chilanga, Zambia, 13.V.1962, A. Angus (M.1957); 2 & 3) PDD 32878, 32879, on dead leaves of *Beilschmiedia tawa,* Orere Pt., Manukau Co., N.Z., 3.I.1974, B. Kendrick (KNZ 116, 159); 4) PDD 32880, on leaves of *B. tawa,* Hongi's Tr. Scenic Reserve, Hwy. 30, N.Z., 12.I.1974, B.K. (KNZ 246).

Known distribution: New Zealand, Zambia.

   C. africana differs from *C. bulbosa* in its narrowly ampulliform and smaller phialides with conic or subcylindrical venters, and in its smaller conidia. It differs from *C. setosa* in its non-septate setae and its smaller, unicellular conidia.

2) Chaetochalara aspera Pirozynski & Hodges (Figure 48)
in Can. J. Bot., 51: 157, 1973.

   Colony hypophyllous, superficial, effuse, brown to brownish black, hairy or setose, covered by a white, downy mass of conidia. Fertile hyphae superficial, subhyaline to pale brown, septate and branched, bearing setae and phialophores. Setae simple, erect, subcylindric to subulate, multiseptate, infrequently constricted at the septa, dark brown to smoky brown below, paler above, occasionally con-colorous with the phialophores; wall smooth, 1μ thick; 79-285 [210] μ long, 5.5-6.5 [6] μ wide at the base, and 2.5-4.5 [3.2] μ wide at the apex. Phialophores simple, erect, cylindrical, up to 4-septate or occasionally many-septate, dark brown below, paler above, 48-140 [84] μ long, slightly expanded at the base to a width of 5-6.5 μ, and terminating in a phialide. Phialides ampulliform, pale brown to brown, 48-77 [64] μ long; venter 18-32 [23] μ long, narrow at the base and expanded at its median part to a width of 5-7 [6.2] μ; collarette cylindrical, 30-47 [41] μ long, 3-5 [4] μ wide; transition from venter to collarette gradual, occasionally marked by a constriction; collarette wall sparsely asperate. Phialoconidia extruded singly or in short chains, cylindrical, rounded at the apex and truncate at the base with a barely perceptible marginal frill, mostly 1-septate, occasionally 2-septate, cells unequal; hyaline, wall smooth and not constricted at the septa; 10-27 [19] x 2.5-4 [3.3] μ; mean conidium length/width ratio = 6:1.

Habitat: On dead leaves of *Knightia excelsa, Myrica cerifera, Persea borbonia.*

Specimens examined: 1 & 2) DAOM 127892, 139268, on leaves of *Myrica cerifera,* Research Triangle Pk., N.C., U.S.A., C. S. Hodges; 3) DAOM 137841 [*Holotype*], on leaves of *Persea borbonia,* nr. Aiken, S.C., U.S.A., 7.X.1971, C. S. Hodges; 4) PDD 32877, on leaf of *Knightia excelsa,* Kauaeranga Valley, Thames Co., N.Z., 21.I.1974, B. Kendrick (KNZ 213).

Known Distribution: New Zealand, U.S.A.

   C. aspera is distinct from all other known species of *Chaetochalara* in possessing stalked phialides with asperate collarettes, and in its larger phialides and conidia.

   The holotype specimen and the New Zealand collection both bear a discomy-cete intimately associated with *C. aspera,* an association that suggests possible rela-tionship between the two. A full description of the discomycete is given on page 183.

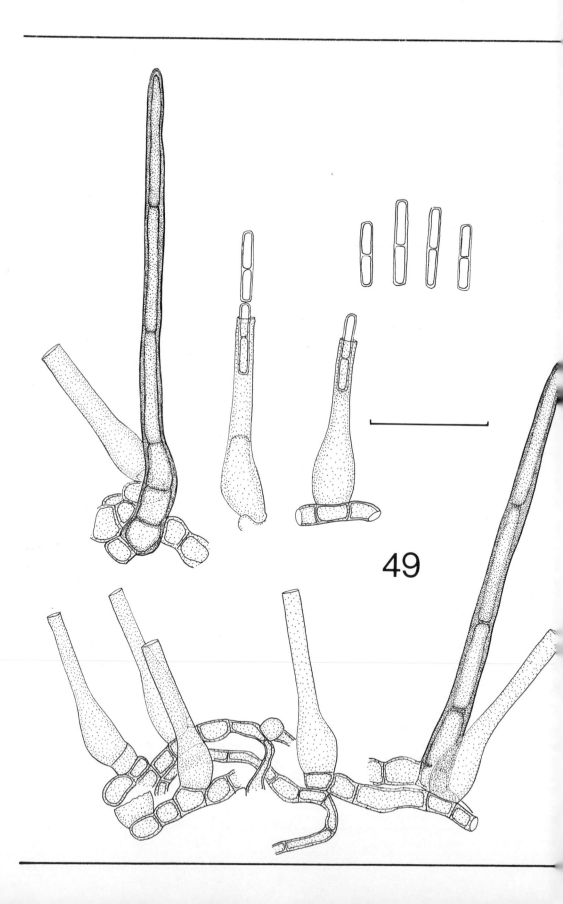

49

3) **Chaetochalara bulbosa** Sutton & Pirozynski (Figure 47B)
in Trans. Br. mycol. Soc., 48: 351, 1965.

Colonies amphigenous, mostly hypophyllous, effuse, dark brown to black, hairy, scattered in irregular patches. Fertile hyphae superficial, hyaline to pale brown, septate and branched, bearing phialides and setae. Setae simple, erect, 50-90 [72] μ long, expanded at the base into a usually bulbous or occasionally conical vesicle 7.5 μ long and 5.5-7.5 [6.7] μ wide, gradually narrowing toward a blunt or pointed apex 1-2 [1.5] μ wide; non-septate, dark brown above, paler below, wall smooth. Phialides arising directly from the superficial hyphae, ampulliform, pale brown or yellowish brown, 16-54 [37] μ long; venter bulbous, with a maximum median width of 5-7 (-9) [6.5] μ, narrowing at a height of 7.5-10 μ to a cylindrical collarette 1.5-2.5 [2.2] μ wide; wall smooth; transition from venter to collarette abrupt. Phialoconidia extruded in easily dispersible chains, cylindrical with rounded or blunt ends, unicellular, hyaline, smooth-walled, 6.5-10 [8.7] x 1-2 [1.5] μ; mean conidium length/width ratio = 5.3:1.

Habitat: On rotting leaves of *Ilex aquifolium.*

Specimens examined: 1) IMI 89645(a) [*Holotype*], Studland, Dorset, U.K., 26.V.1962; 2) IMI 86894(c) [*Paratype*]; 3) DAOM 126877 ex WINF(M) 9499(a), Box Hill, Surrey, U.K., 15.V.1968, B. C. Sutton.

Known distribution: U.K.

C. bulbosa resembles *C. setosa* in the morphology of its phialides, but can be distinguished by its non-septate and shorter setae with bulbous base, and its smaller, non-septate conidia.

4) **Chaetochalara cladii** Sutton & Pirozynski (Figure 50)
in Trans. Br. mycol. Soc., 48: 352, 1965.

Colonies amphigenous, effuse, dark brown, velvety. Fertile hyphae superficial, pale brown, septate and branched. Setae simple, erect, subulate, usually 3-5-septate, dark reddish brown below, becoming paler above, smooth-walled, 80-155 (-350) [130] μ long, 5-6.5 [5.8] μ wide at the base, tapering toward the apex to a width of 1-2.5 [2] μ. Phialides borne directly on the fertile hyphae or on short basal cells from which sterile, hyaline or pale brown hyphal elements often arise; subcylindrical to obclavate, subhyaline to pale brown, 30-45 [39] x 4.5-9 [6.7] μ, usually about 8 μ wide at the broadest point near the base, narrowing slightly above to 4.5-7 [5.6] μ wide; wall smooth. Phialoconidia produced in chains; cylindrical, rounded at apex and truncate at the base with a marginal frill; unicellular or 1-septate, hyaline; wall smooth and up to 0.5 μ thick; 10-18 [13] x 4-5 [4.3] μ; mean conidium length/width ratio = 3:1.

Habitat: On dead leaves of *Cladium mariscus.*

Specimen examined: IMI 89626(b) [*Holotype*], Sugar Hill, Wareham, Dorset, U.K., 26.V.1961.

Known distribution: U.K.

C. cladii approaches *C. ramosa* in phialide morphology, but its phialides do not arise as branches of short phialophores, and its conidia are larger.

The type specimen now has few conidial fructifications, but bears abundant

Chaetochalara setosa. Setae, phialophores and conidia ex type in CAS.

apothecia of a discomycete which appear to arise from the same mycelium as the conidial fructification, suggesting a possible relationship between the two. A detailed description of the discomycete is given on page 183.

5) **Chaetochalara ramosa** sp. nov. (Figure 5l)

Coloniae hypophyllae, effusae, atrae, caespitosae vel velutinae; albida et pulveracea massa conidii velatae. Phialides et setae super tenui, superficiali strato isodiametrarum, incrassatarum, subhyalinarum, cellularum portatae. Setae paucae, mar-

50)   *Chaetochalara cladii.* Setae, phialophores and conidia ex IMI 89626b.

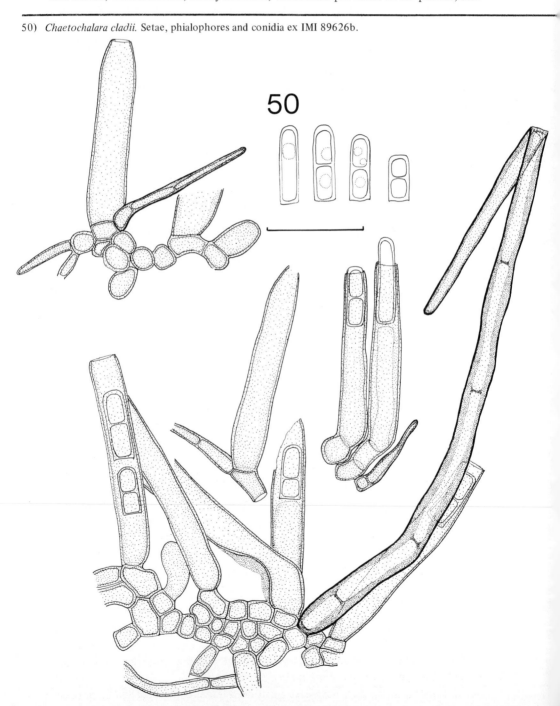

50

ginales, simplices, rectae vel flexae, 140-180 [165] μ long., 3.5-6.5 [5] μ lat. ad basem, versus apicem gradatim decrescentes in 2.5-3 [2.6] μ lat.; 6-9-septatae, sucinaceae, pariete laevi, 1 μ cr. Phialides visae ut rami brevium phialophororum e cellulis basalibus orientes; subcylindraceae, pallide brunneae, flavide-brunneae vel fumoso-brunneae, 25-38 [32] μ long., 3-4.5 [3.6] μ lat., pariete laevi. Phialoconidia in catenas extrusa, cylindracea, apice rotundato, base truncata, fimbriam marginalem ferente, uniseptata, hyalina, pariete laevi et 0.5 μ cr.; 9-14 [11] x 3-4.5 [3.7] μ; ratio conidii long./lat. = 3:1.

Colonies hypophyllous, effuse, black, caespitose to hairy, covered by a white powdery mass of conidia. Phialides and setae borne on a thin, superficial layer of isodiametric, thick-walled, subhyaline, basal cells 3.5-4.5 μ diam. Setae few, marginal, simple, erect or bent, 6-9-septate, amber coloured; 140-180 [165] μ long, 3.5-6.5 [5] μ wide at the base, gradually tapering toward the apex to a width of 2.5-3 [2.6] μ; wall smooth and 1 μ thick. Phialides arising as branches of short phialophores originating from the basal cells; subcylindrical, pale brown, yellowish brown or smoky brown, 25-38 [32] μ long, 3-4.5 [3.6] μ wide at the apex; wall smooth. Phialoconidia in chains; cylindrical, rounded at the apex, base truncate with a marginal frill, 1-septate, hyaline; wall smooth and 0.5 μ thick; 9-14 [11] x 3-4.5 [3.7] μ; mean conidium length-width ratio = 3:1.

Habitat: On leaves of *Dalbergia lactea.*

Specimens examined: 1) IMI 106429(e) [*Holotype*], Tanganyika [Tanzania], 26.I.1964, K. Pirozynski (M 330d); 2) IMI 106429(d) [*Paratype*].

Known distribution: Tanzania.

For affinities of this species refer to *C. cladii.*

6) Chaetochalara setosa (Harkn.) comb. nov. (Figure 49)
≡ *Chalara setosa* Harkn.
in Calif. Acad. Sci. Bull. 3: 164, 1885.

Colonies amphigenous, effuse, black, velutinous to hairy, in irregular patches. Fertile hyphae superficial, septate, branched, bearing setae and phialides. Setae simple, erect, 4-6-septate, dark brown to reddish brown, 80-140 [110] μ long, slightly inflated at the base to a width of up to 6.5 μ, gradually tapering toward the apex to a width of 2.5-3.5 [3.1] μ; wall smooth and 1.5 μ thick. Phialides borne directly on the fertile hyphae, ampulliform, subhyaline to pale brown, 25-38 [31] μ long, with a bulbous venter 9-13 [11] μ long and 5.5-7.5 [7] μ wide just above the base, and a cylindrical collarette 16-25 [20] μ long and 2.5-3.5 [3] μ wide; transition from venter to collarette abrupt. Phialoconidia in chains; cylindrical, rounded at the apex and truncate at the base, 1-septate, hyaline; wall smooth and not constricted at the septum; 11-15 [13] x 1.5-2.5 [2] μ; mean conidium length/width ratio = 5.3:1.

Habitat: On leaves of *Quercus densiflora.*

Specimen examined: CAS 2907 [*Holotype*], Tamalpais, Calif., U.S.A., XI.1881, Harkness.

Harkness's description of the fungus is slightly at variance with the above, in that he characterized the setae as non-septate and gave the dimensions of phialides and conidia as 12 x 24 μ and 4 x 20 μ respectively.

Affinities of this species have been mentioned under *C. africana* and *C. bulbosa.*

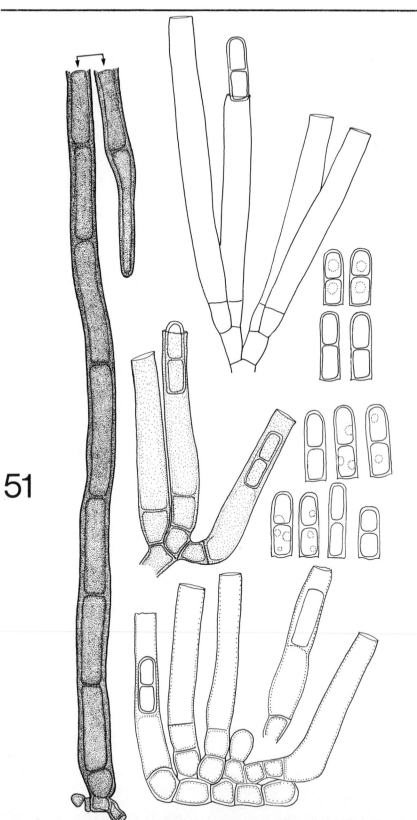

51

Sporoschisma Berkeley & Broome apud Berkeley*
in Gardener's Chronicle, p. 540, 1847 (footnote); Hughes in Mycol. Pap., 31: 2, 1949; Hughes, in N.Z. Jl. Bot., 4: 78, 1966.

Colonies discrete, superficial, effuse, black, hairy to setose, composed of phialophores and capitate hyphae arising from small stromata. Capitate hyphae erect, sterile, with an apical swelling surrounded by a blob of (?) mucilage. Phialophores subcylindrical to clavate, septate in the basal part, brown to dark brown, terminating in a phialide. Phialides subcylindrical to urceolate with long collarettes. Phialoconidia more or less cylindrical, ends flattened or rounded, several-septate, pale brown to dark brown, end cells paler than central cells, wall smooth or finely roughened.

Type species: Sporoschisma mirabile Berk. & Br.

Key to Species
1) Conidia 3-septate 2
1) Conidia 5-septate 3

   2) Conidia pale brown, wall rough; 7.5-11 μ
      wide . . . . **Sporoschisma juvenile** (1)
   2) Conidia dark brown, especially at the septa and extremities, wall smooth;
      10-14 μ wide . . . . **Sporoschisma mirabile** (2)

   3) Phialoconidia with the 3 central septa obscured by wide bands of dark pigmentation; 12-14 μ wide . . . . **Sporoschisma nigroseptatum** (3)
   3) Phialoconidia with septa not obscured by wide bands of dark pigmentation; 9-12 μ wide . . . . **Sporoschisma saccardoi** (4)

1) Sporoschisma juvenile Boudier (Figure 52A)
in Icones Mycologicae, Explication des Planches, p. 12, 1904; Icones Mycologicae, tome IV, Texte descriptif, p. 348, ?1911; Hughes, Mycol. Pap., 31:15, 1949.

Colony superficial, effuse, black, hairy to loosely caespitose. Capitate hyphae intermixed with phialophores, scattered or in tufts; 100-160 μ long, 2-5-septate, thick-walled and dark brown below, becoming thin-walled and hyaline at the apex; 5-8 μ wide at the base, 4-5.5 μ wide below the swollen, bulbous apex which is 5.5-8 μ wide and invested by a blob of (?) mucilage. Phialophores solitary and scattered, or in sparse tufts, erect, straight or variously bent, subcylindrical, up to 240 μ long, 8-10 μ wide just above the bulbous base, 1-3-septate in the basal part, dark brown; terminating in a phialide. Phialides lageniform to urceolate, slightly rough-walled, 90-170 [125] μ long, composed of a subcylindrical to ellipsoidal venter 35-55 [43] x 14-21 [17] μ, and a cylindrical collarette 52-115 [81] μ long and 9.5-12 [11] μ wide; transition from venter to collarette abrupt; ratio of mean lengths of collarette and venter = 1.8:1. Phialoconidia cylindrical with rounded ends, 3-septate, pale brown, rough-walled, 20-42 [31] x 7.5-11 [9] μ; mean conidium length/width ratio = 3.4:1.

Habitat: On wood of *Alnus, Fagus, Fraxinus, Hedera, Quercus* and undetermined trees.

*   In using this nomenclator, we are following Hughes (1949).

Chaetochalara ramosa. Setae, phialophores and conidia ex IMI 106429e.

Specimen examined: DAOM 34481, on *Fagus sylvatica* wood, Ranmore Common, Surrey, U.K., 13.VI.1948 (ex IMI 29345 [e]), (authentic for the name).

Known distribution: Canada, France, U.K.

S. juvenile is close to *S. mirabile* but can be distinguished by its (a) paler, narrower conidia with rounded ends and rough walls, (b) narrower collarettes, (c) more tufted phialophores, and (d) shorter capitate hyphae.

2) Sporoschisma mirabile Berk. & Br. apud Berk. (Figure 52B)
in Gardener's Chronicle, p. 540, 1847 (footnote); Hughes, Mycol. Pap., 31:5, 1949; Hughes, N.Z. Jl. Bot., 4:78, 1966.

Colonies effuse, superficial, black, hairy. Capitate hyphae mixed with phialo-

52A) *Sporoschisma juvenile.* Capitate hyphae, phialophore and conidia ex DAOM 34481.

52B) *Sporoschisma mirabile.* Capitate hyphae, phialophore and conidia ex DAOM 44806.

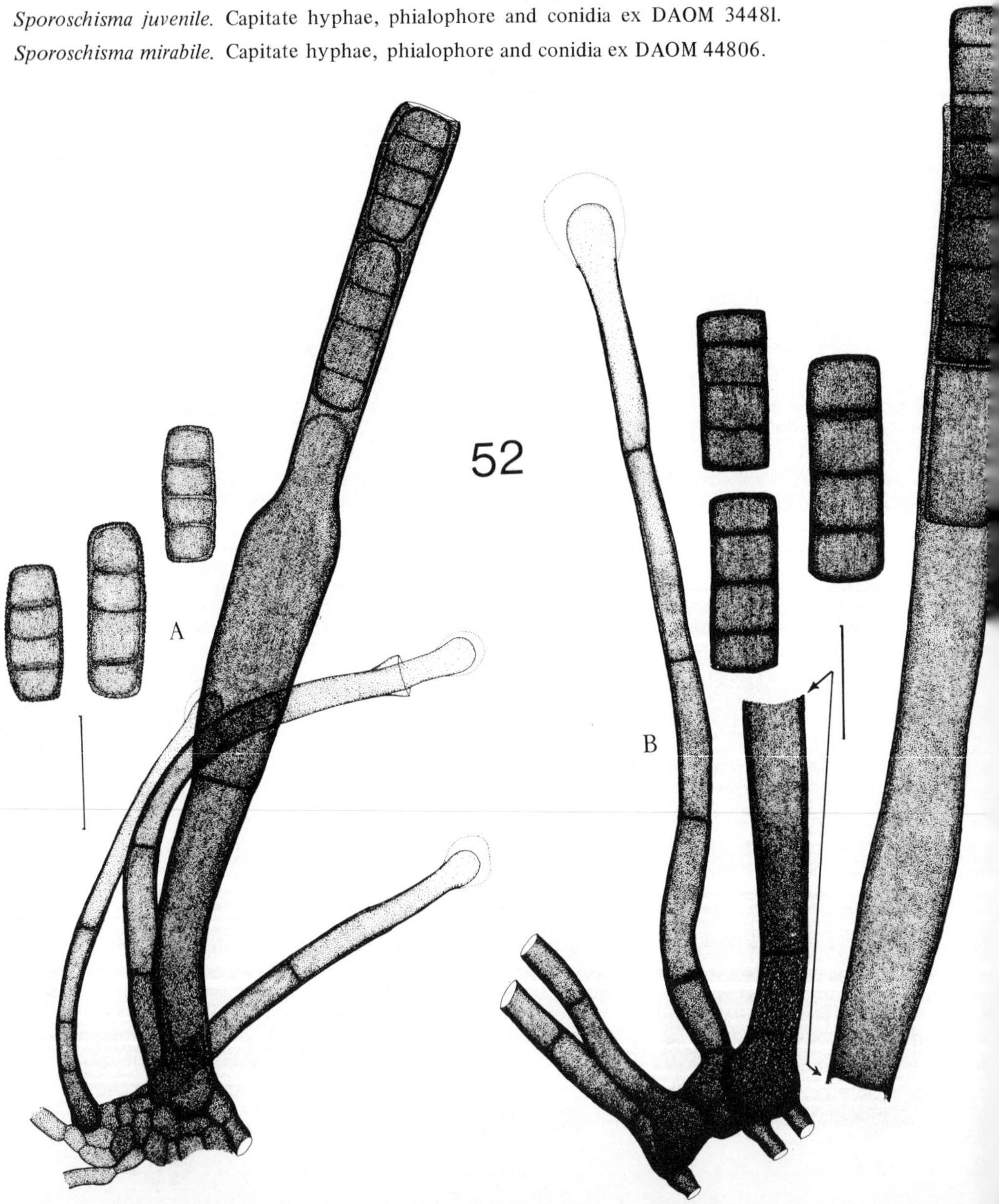

52

A) *Sporoschisma nigroseptatum.* Capitate hyphae, phialophore and conidia ex DAOM 93761a.

B) *Sporoschisma saccardoi.* Capitate hypha, phialophore and conidia ex DAOM 109630.

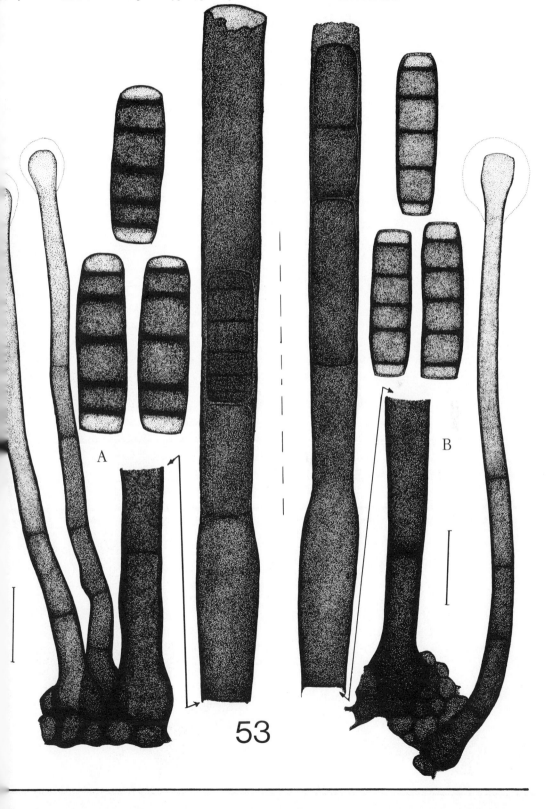

53

phores, 90-250 μ long, 5.5-10 μ wide at the base, 2-5-septate and dark brown below, continuous and paler above, 4-6 μ wide just below the swollen, club-shaped, hyaline apex which is 5.5-9 μ wide and often carries a thick sheath of (?) mucilage. Phialophores solitary, or in groups of up to 8, originating from a thin, erumpent stroma, erect or slightly bent, cylindrical, 160-320 μ long and 6-11 μ wide for the most part except at the bulbous base; 1-3-septate, dark brown and thick-walled in the basal part, terminating apically in a phialide. Phialides subcylindrical, up to 190 μ long and composed of a more or less cylindrical venter 14-20 μ wide, and a narrower, cylindrical collarette 12-16 μ wide, becoming paler toward the apex; transition from venter to collarette gradual. Phialoconidia cylindrical with flattened ends, 3-septate, dark brown, especially around the septa and extremities, smooth-walled; 23-46 [34] x 10-14 [12] μ; mean conidium length/width ratio = 2.8:1.

Habitat: On wood of *Alnus, Bambusa, Beilschmiedia, Corynocarpus, Dysoxylum, Epilobium, Fagus, Fraxinus, Freycinetia, Knightia, Melicytus, Neopanax, Quercus, Rhipogonum, Rhapalostylis, Salix, Sorbus,* and *Ulmus.*

Specimen examined: DAOM 44806, on *Fagus sylvatica,* Fontainebleau, France. VII. 1954, S. J. Hughes (authentic for the name).

Known distribution: Belgium, Canada, France, Germany, New Zealand, U.K., U.S.A.

Affinities of this species are mentioned under *S. juvenile. S. mirabile* is considered to be the conidial state of *Melanochaeta aotearoae* (Hughes) Müller, Harr et Sulmont.

3) Sporoschisma nigroseptatum Rao & Rao (Figure 53A)
in Mycopath. Mycol. appl., 24: 82, 1964; Hughes, N.Z. Jl. Bot., 4: 81, 1966.

Colony sparse, effuse, black, hairy. Capitate hyphae and phialophores arising from a superficial, brown to dark brown, pulvinate, pseudoparenchymatous stroma. Capitate hyphae in tufts of up to 15, erect or slightly bent, subulate with a swollen basal cell; 80-160 μ long, 2-6-septate and dark brown below; the apical cell is 45-90 μ long and gradually tapers to a width of 4.5-5 μ below the subhyaline, swollen apex which is up to 9 μ wide and bears a hyaline cap of (?) mucilage. Phialophores in tufts of up to four; erect, cylindrical to subcylindrical, 217-310 μ long, 9-11 μ wide just above the bulbous base, but broader above, 1-3-septate and brown to dark brown; terminating in a phialide. Phialides subcylindrical, 165-230 [205] μ long, composed of an obconical venter 40-55 [47] μ long and up to 18 μ wide, and a subcylindrical collarette 135-180 [160] μ long which gradually widens to 16-20 μ at the apex; transition from venter to collarette abrupt; ratio of mean lengths of collarette and venter = 3.4:1. Phialoconidia in short, fragile chains; cylindrical with flattened or slightly rounded ends, unequally 5-septate, median cells dark brown, end cells subhyaline, wall smooth and constricted at the septa, the three central septa obscured by darkly pigmented bands of wall about 3.5 μ wide; 34-47 [40] x 12-14 [13] μ; mean conidium length/width ratio = 3:1.

Habitat: On *Cortaderia, Saccharum.*

Specimen examined: DAOM 93761(a), on *Cortaderia atacamensis,* Upper Piha Valley, Waitakere Ranges, Auckland Prov., N.Z., 9.X.1963, S. J. Hughes (1500a) (authentic for the name).

Known distribution: India, New Zealand.

The first-formed phialoconidia differ from others in that they are shorter, more rounded at the apex, and have fewer septa. Occasionally the mature conidia are non-septate or variably septate.

S. nigroseptatum is close to *S. saccardoi,* but can be distinguished by its (a) darker and larger phialophores, (b) larger conidia with wide bands of very dark wall masking the three central septa, and (c) larger central cells.

4) Sporoschisma saccardoi Mason & Hughes apud Hughes (Figure 53B)
in Mycol. Pap., 31: 20, 1949; Hughes, N.Z. Jl. Bot., 4: 84, 1966.

Colonies sparse, superficial, dark brown to black, velutinous. Capitate hyphae mixed with phialophores; subulate, erect, straight or flexuous, up to $150\,\mu$ long, 5-6.5 μ wide at the base, and 4-5.5 μ wide just below the swollen apex which is 6-8 μ wide; 2-4-septate and brown below, becoming paler above and subhyaline at the apex which is enveloped in (?) mucilage. Phialophores solitary and scattered, mixed with capitate hyphae; erect, straight or slightly bent, more or less cylindrical, dark brown at the base, paler above, up to $210\,\mu$ long, swollen at the base to a width of 10-16 μ, 1-septate; stalk cell 7-10 μ wide; smooth-walled, terminating in a phialide. Phialides subcylindrical to lageniform, 180-210 [195] μ long, composed of a narrowly obconical or subcylindrical venter 37-50 [45] x 15-19 [17] μ and an obconical collarette 140-160 [150] μ long and 12-15 [13] μ wide at the apex; transition from venter to collarette usually gradual, rarely abrupt; ratio of mean lengths of collarette and venter = 3.3:1. Phialoconidia in short chains; cylindrical, flattened at the ends, 5-septate, median cells dark brown, end cells lighter, wall smooth and not constricted at septa; 32-48 [40] x 9-12 [11] μ; mean conidium length/width ratio = 3.6:1.

Habitat: On wood of *Pyrus, Salix.*

Specimens examined: 1) DAOM 109630 (slide) ex Saccardo, Mycotheca Veneta 1586, on wood of *Pyrus malus,* Italy, X.1880 (in NY), (authentic for the name); 2) on wood, Queen's Univ. Biol. Sta., Lake Opinicon, Ont., Canada, 10.IX.1966, C. J. Wang.

Known distribution: Canada, Indonesia, Italy.

Affinities of this species are indicated under *S. nigroseptatum. S. saccardoi* is the presumed conidial state of *Melanochaeta hemipsila* (Berk. & Br.) Müller, Harr et Sulmont.

Sporendocladia Arnaud ex Nag Raj & Kendrick
in Bull. Soc. mycol. Fr., 69: 279, 1953.

Cellulae conidiogenae (phialides) ad apicem simplices vel ramosi conidiophorii in fasciculos radiantibus dispositae; phialides et conidia ut in *Chalara*.

Species typica: Sporendocladia castaneae Arnaud.

Sporendocladia castaneae Arnaud ex Nag Raj & Kendrick (Figure 54)
in Bull. Soc. mycol. Fr., 69: 279, 1953.

Colonia superficialis, effusa, velutina, brunnea. Phialophora solitaria vel aggregata, recta vel varie flexa, simplicia aut ramosa, septata, 60-165 [125] μ long., base 5-6 μ lat., ad basem brunnea et ad apicem pallescentia in fasciculos phialidum 6-8 terminantia. Phialides lageniformes, 11-15 (-18) [13] μ long.; venter globosus vel subglobosus, 6-8.5 (-10) [7.6] x 2-4 [3.4] μ; collum cylindraceum, 4.5-7.5 [6] x 1.5-2 [1.6] μ; transitio ex ventro ad collum abrupta vel gradatim; ratio long. colli et ventris = 1.3:1. Phialoconidia in catenas extrusa; enteroblastica, cylindracea, extremis truncatis, unicellularia, hyalina, pariete laevi, 2-3 [2.6] x 1-1.5 μ.

Habitat: in cupulas *Castaneae vescae.*

Lectotypus: PC, sub numero 2230, IX.1945, G. Arnaud.

) *Sporendocladia castaneae.* Phialophore and conidia ex type in PC.

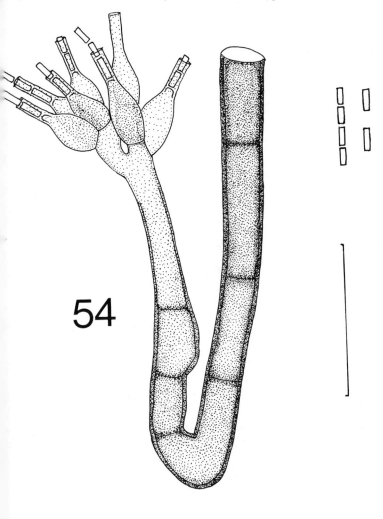

54

Ascoconidium Seaver

in Mycologia, 34: 414, 1942; Funk in Can. J. Bot., 44: 39, 1966.

Phialophores arising from a pseudoparenchymatous basal stroma and becoming erumpent through the periderm, simple, short, sparsely septate, terminating in a phialide. Phialides aggregated in densely packed sori; clavate to subcylindrical, brown, thick-walled. Conidia cylindrical, septate, hyaline, formed singly and successively.

Type species: Ascoconidium castaneae Seaver.

Ascoconidium is distinct from *Chalara* and *Sporoschisma* because: (a) its phialides lack a morphogically well-differentiated venter; (b) the conidiogenous locus is near the base of the phialide; (c) the walls of the phialide are thick, and rupture at maturity by a vertical split at the apex, the lacerated flaps seemingly falling back into place after the release of each conidium.

Key to Species
1) Conidia consistently 3-septate (phragmoconidia), 30-35 (-40) x 8-10
    μ . . . . **Ascoconidium castaneae** (1)
1) Conidia 3-7-septate (often dictyoconidia), 40-60 x 10-14
    μ . . . . **Ascoconidium tsugae** (2)

1) Ascoconidium castaneae Seaver (Figure 55)
in Mycologia, 34: 414, 1942; Funk in Can. J. Bot., 44: 39, 1966.

Colony arising as a discrete sorus from a pseudoparenchymatous basal stroma. Phialophores becoming erumpent through the periderm, and arranged in a densely packed palisade; simple, short, narrow, 1-septate, brown, terminating in a phialide. Phialides clavate or subcylindrical with a rounded apex, concolorous with the phialophore, 75-110 [95] μ long, 9.5-15 μ wide at the base and 12-19 [16] μ wide bearing a conspicuous marginal frill; 3-septate, hyaline, thick-walled, 30-35 (-40) x 8-10 μ.

Habitat: On *Castanea dentata.*

Holotype: NY - N. Amer. Fungi 2147, on dead limbs of *Castanea dentata*, Winchester, Pa., U.S.A., VIII.1888.

Known distribution: U.S.A.

2) Ascoconidium tsugae Funk (Figure 56)
in Can. J. Bot., 44: 219, 1966.

Colony black, erumpent. Phialophores arising from a dark brown, pseudoparenchymatous basal stroma and arranged in a densely packed palisade; simple, short, narrow, cylindrical, up to 4-septate, dark brown, terminating in a phialide. Phialides clavate or subcylindrical with a rounded apex, concolorous with the conidiophore, 75-110 [95] μ long, 9.5-15 μ wide at the base and 12-19 [16] μ wide at the broadest point; wall thick and somewhat verrucose. Phialoconidia formed singly; cylindrical, with a rounded apex and a truncate base bearing a conspicuous marginal frill; 3-7-septate; often dictyoconidia with 1 to 3 vertical septa; hyaline, thick- and smooth-walled; 40-60 [51] x 10-14 [13] μ.

Habitat: On *Tsuga heterophylla.*

Specimens examined: 1) DAVFP 15669 [*Holotype*], Cowichan Lake, Vancouver Is., B.C., Canada, 12.XII.1963, A. Funk and Y. Hiratsuka (*Isotype:* DAOM 105793); 2) DAOM 57868, Courtenay, Vancouver Is., B.C., Canada, A. T. Foster.

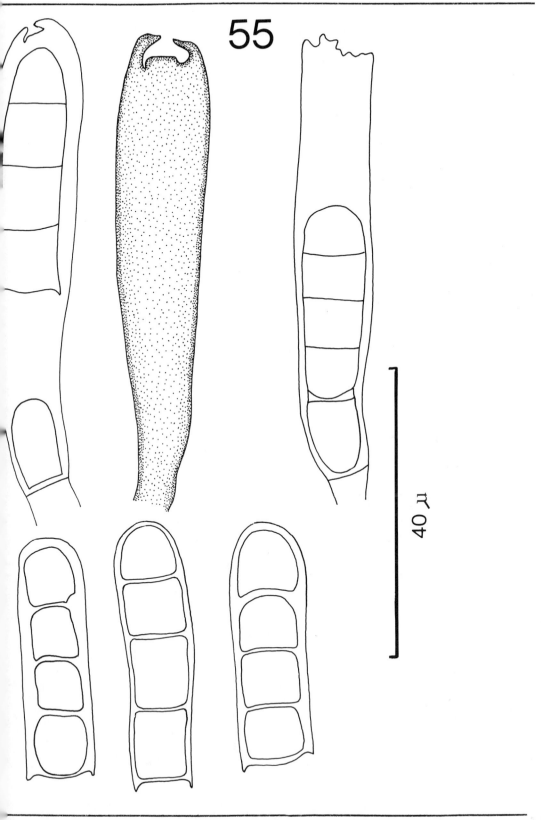

Ascoconidium castaneae. Phialides and conidia [after Funk. 1966].

Known distribution: Canada.

This species differs from *A. castaneae* in possessing larger phialides, and larger conidia that are 3-7-septate and often muriform.

56) *Ascoconidium tsugae.* Phialides and conidia ex DAVFP 16559.

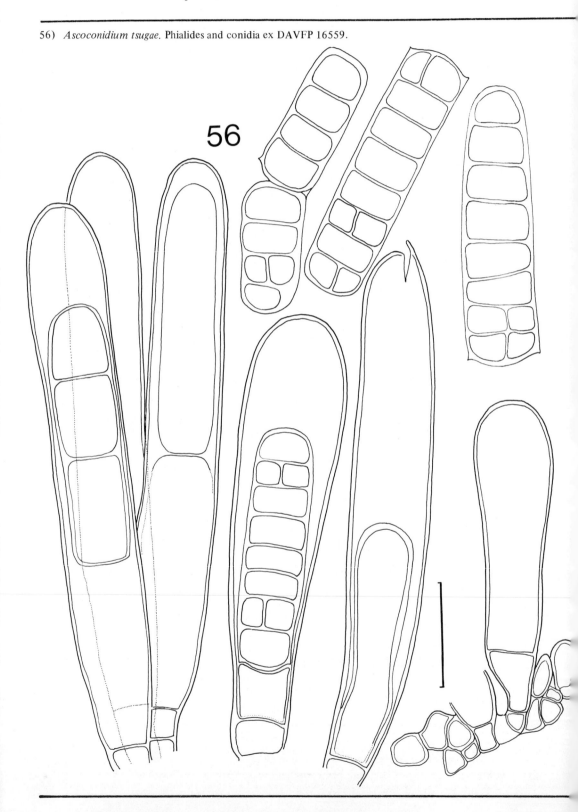

56

Bloxamia Berk. and Br.

in Ann. Mag. nat. Hist., 13: 468, 1854.

= *Endosporostilbe* Subram.

in J. Indian bot. Soc., 37: 47, 1958.

Phialophores densely aggregated in a sporodochium or synnema arising from a basal stroma; erect, cylindrical, septate, subhyaline, pale brown or brown, terminating in a phialide. Phialides cylindrical to subcylindrical with collarette not visibly differentiated from the venter. Phialoconidia cuboid or short cylindrical, oblong with truncate ends, unicellular, hyaline to subhyaline.

Type species: Bloxamia truncata Berk. and Br.

Bloxamia shares its enteroblastic-phialidic mode of conidium ontogeny with *Chalara* and *Sporoschisma,* but is distinct because its phialophores are densely aggregated into sporodochia or synnemata, and the collarette is not visibly differentiated from the venter of the phialide. It differs from *Ascoconidium* in its phialophores, which are not in lax clusters, and in its conidia, which are unicellular and thin-walled.

Key to Species

1) Fructification a sporodochium; phialophores arising from a superficial basal stroma; phialoconidia short-cylindrical, oblong, hyaline to subhyaline, 2-4 (-7.5) x 1.5-2.5 μ . . . . **Bloxamia truncata** (2)

1) Fructification a synnema; phialoconidia cylindrical, hyaline, 3-6.5 x 2-3 μ . . . . **Bloxamia nilagirica** (1)

1) Bloxamia nilagirica (Subram). comb. nov. (Figure 7)

≡ *Endosporostilbe nilagirica* Subram.

in J. Indian Bot. Soc., 37: 49, 1958.

Synnemata conspicuous, scattered, subcylindrical, erect, brown, 1120-1260 μ long, 260-380 μ wide at the base, 140-220 μ wide at the apex, composed of densely aggregated and coherent conidiophores. Phialophores septate, 1-2 μ wide at the base, gradually increasing in width, subhyaline but becoming darker brown and loosely packed toward the distal end, thin-walled, terminating in a cylindrical or subcylindrical phialide. Phialoconidia occurring in long chains that cohere in a slimy mass over the synnematal head; cylindrical with blunt or truncate ends, unicellular, hyaline, thin- and smooth-walled, 3-6.5 x 2-3 μ; mean conidium length/width ratio = 1.9:1.

Habitat: On twigs of undetermined trees.

Specimen examined: Slide ex Madras Univ. Bot. Lab. #1926 [*Holotype*] , on twigs of trees among litter in Government Gdns., Ootacamund, Madras State, India, 23.XI.1957, C. V. Subramanian.

Known distribution: India.

Bloxamia nilagirica differs from *B. truncata* in the aggregation of its phialophores to form synnemata, and in its hyaline, cylindrical, and slightly larger phialoconidia.

2) Bloxamia truncata Berk. and Br. (Figure 57)

in Ann. Mag. nat. Hist., 13: 468, 1854.

= *Hormococcus nitidulus* Sacc.

in Michelia, 2: 285, 1881, fide Pirozynski & Morgan-Jones,

in Trans. Br. Mycol. Soc., 51: 185, 1968.

= *Trullula nitidula* (Sacc.) Sacc.
in Sylloge Fung., 3: 72, 1886, fide Pirozynski & Morgan-Jones,
in Trans. Br. Mycol. Soc., 51: 185, 1968.
= *Bloxamia nitidula* (Sacc.) Höhn.
in Annls. mycol., 1: 405, 1903, fide Pirozynski & Morgan-Jones,
in Trans. Br. mycol. Soc., 51: 185, 1968.
= *Bloxamia saccardiana* Allesch.

57) *Bloxamia truncata.* Phialides and conidia. A. ex FH 168; B. ex FH 3839/1902; C. ex FH 3839/1903; D. ex FH 472; E. ex Curtis herb. sub *B. tetraploa*; F. ex FH, sub *B. leucophthalma*.

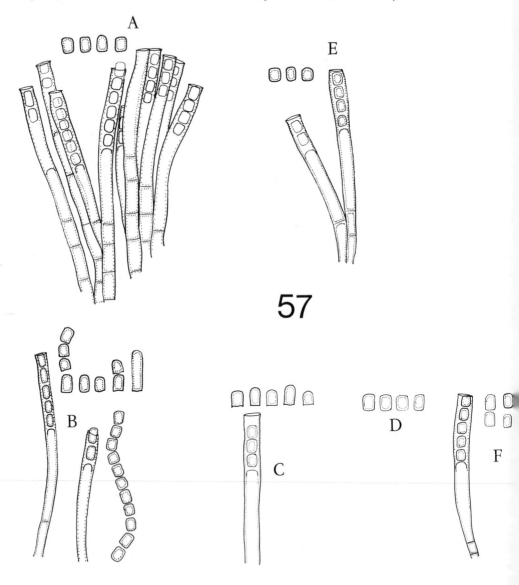

in Rabenh. Kryptogamenflora, 7: 553, 1903, fide Pirozynski & Morgan-Jones,
in Trans. Br. mycol. Soc., 51: 185, 1968.
= *Bloxamia leucophthalma* (Lev.) Höhn.
in Sber. Akad. Wiss. Wien, Abt. I, 119: 37, 1910.

Sporodochia scattered, or gregarious and confluent, disciform, black, 140-180 (-500) μ diam. Phialophores arising from upper cells of a superficial stroma composed of long, loosely aggregated cells, arranged in a densely packed palisade layer; erect, cylindrical, septate, subhyaline to pale brown, smooth-walled, terminating in a phialide. Phialides cylindrical to subcylindrical, 15-32 [26] x 2-3 [2.5] μ. Proliferation occasional, percurrent. Phialoconidia occurring singly or in easily dispersible chains; short-cylindrical to oblong, with a rounded apex and a truncate base often adorned with an inconspicuous marginal frill, or both ends somewhat obtuse; unicellular, hyaline to subhyaline, smooth-walled, 2-4 (-7) [2.8] x 1.5-2.5 [2] μ; mean conidium length/width ratio = 1.5:1.

Habitat: On *Fagus, Salix, Ulmus.*

Specimens examined: 1) K [*Holotype*], on wood of *Ulmus montana,* Bath, Somerset, U.K., 5.III.1852, C. E. Broome; 2) FH-3839, Herb. Höhnel, on *Salix,* Jaize, Bosnien [Yugoslavia], IV.1903, Höhnel; 3) FH 3839, Herb. Höhnel on *Fagus,* Fressbauen, Wilhelmstrasse, 13.VII.1902, Höhnel; 4) FH 3839, slide ex Herb. Höhnel, sub *Bloxamia leucophthalma* (Lev.) Höhn., *Ulmus?* Bois de Boulogne, Paris, France, Höhnel; 5) FH, sub *Bloxamia tetraploa* B. and Br., U.K., Berkeley.

Exsiccati examined: FH, M. C. Cooke. Fungi Britannici exsiccati No. 472, Batheaston, Somerset, U.K., C. E. Broome; Rabenh. Fungi Europaei, No. 168 on dry wood of *Ulmus,* Batheaston, Somerset, U.K., II. 1856, C. E. Broome.

Known distribution: France, Germany, U.K., Yugoslavia.

Little Known, Dubious or Excluded Taxa

1) Ceratocystis antennaroidospora Roldan
in Philipp. J. Sci. 91(4): 422, 1962.

Roldan (*op. cit.*) applied the name *Ceratocystis antennaroidospora* Roldan to a fungus isolated from stems of *Calamus maximus* at Los Baños, Philippines. His description and illustrations were of the asexual state only, and he clearly stated that he could not obtain perithecial material. No type was designated, and apparently no type specimen or culture exists. In Roldan's choice of a name for the fungus, a contradiction of the Code is apparent, permitting conflicting interpretations. Under Article 59, this name is to be judged illegitimate although validly published. With this interpretation, application of the species name to anything is excluded since the author misclassified the species or misapplied the generic name. On the other hand, the interpretation implied in the statement, "Some time in future it is hoped that the perithecial structures may be found . . ." (Roldan, *op. cit.*, p. 417), would permit us to invoke article 34, which renders *C. antennaroidospora* a *nomen provisiorum.* We consider this interpretation as more appropriate and the binomial as invalidly published. Groves and Elliott (1969) dealt with the similar case of *Sclerotinia alni* Maul in exactly the same way, though for a different reason.

According to Roldan, the fungus appeared somewhat similar to the conidial state of *Ceratocystis radicicola, C. variospora* and *C. moreans* [*sic*], but had smaller conidia. This distinction seems to be dubious. Since neither the original material nor an authentic collection is available for study, we are unable to determine its specific identity; however, Roldan's illustration is reminiscent of the *Chalara* state of *Ceratocystis radicicola.*

2) Ceratocystis asteroides Roldan
in Philipp. J. Sci. 91(4): 421, 1962.

This binomial, employed for isolates from *Calamus maximus, Cocos nucifera* and *Arenga pinnata* at Los Baños, Philippines, is a misnomer and subject to the same conflicting interpretations as *C. antennaroidospora* (see above). We treat it as invalidly published.

C. asteroides was believed to be similar to the imperfect state of *C. paradoxa,* but for the fact that it produced synnemata. This again is a spurious distinction. From Roldan's illustrations, it is fairly clear that he was dealing with *Chalara paradoxa* (conidial state of *Ceratocystis paradoxa*) which does, in fact, produce synnemata in cultures and is also known to occur on *Cocos nucifera* and *Arenga pinnata* (see substrate index).

3) Chalara ampullula (Sacc.) Sacc. var. **minor.** Sacc. (Figure 58A)
in Ann. Mycol. 9: 255, 1911; Sylloge Fung. 22: 1363, 1925.

Saccardo (1911) described *Chalara ampullula* var. *minor* for a collection made on decayed wood in Lyndonville, N.Y., U.S.A. by Fairman. His description was as follows: "A typo different basi phialoformi minore, nempe 10-12 x 4-4.5 non 14 x 7; conidiis cylindricis utrinque truncatis, 6-8 x 2."

The holotype specimen in PAD has been examined. The colony on the substrate is effuse, superficial, white or creamy white, forming irregular patches, and consists of hyaline, erect, thin-walled, almost imperceptibly septate conidiophores. The conidiophore terminates in a globose or subglobose, vesicular swelling on which the conidia are borne in a head. The conidia are cylindrical with a subtruncate base and a rounded apex; hyaline, 1-septate, 5.5-13 x 2 μ, with thin, smooth walls, and occur in short chains, being connected to each other by very narrow isthmi. No phialides

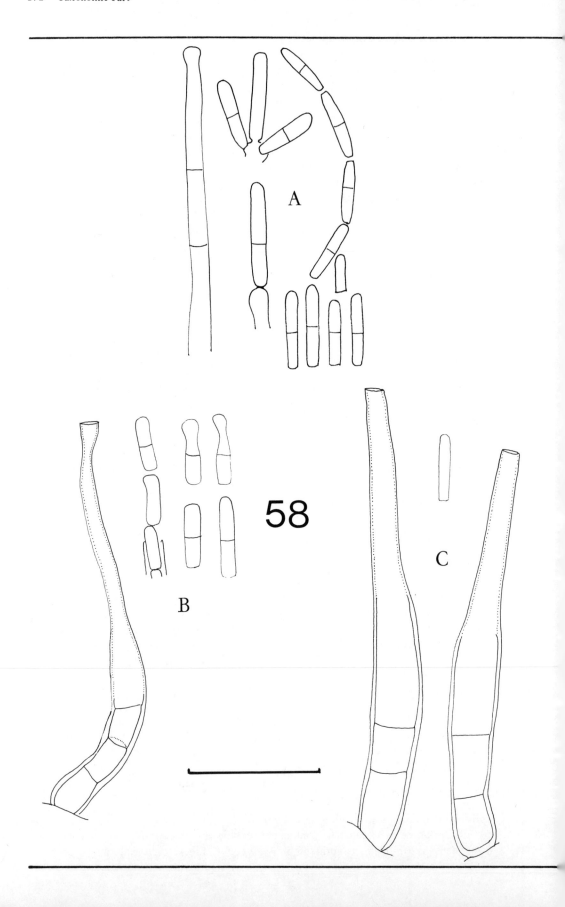

or 'endoconidia' were found on the specimen.

This fungus is excluded from *Chalara,* with which it obviously has no affinity. It may belong in *Hormiactis,* or in the recently published *Anungitea* Sutton (Sutton 1973).

4) Chalara brefeldi Lindau

in Rabenh. Kryptogamenflora, 8: 750, 1907; Sylloge Fung. 22: 1363, 1925.
= *Polyscytalum fungorum* Sacc.
in Sylloge Fung., 4: 336, 1856.

Tulasne and Tulasne (1865) described the conidial form of *Hypomyces asterophorus* Tul. as producing 'oidium'-like, one-celled conidia through basipetal septation and disarticulation of long, thin, alternately or dichotomously branched, tuft-like filaments. Plowright (1884) described the conidia as being formed at the extremities of the branching mycelial tubes. Both these authors noted the presence of chlamydospores, and assigned the fungus to *Hypomyces.* Brefeld (1891), examining cultures of the fungus, described the conidiophores as brown, conical mycelial branches of variable length, bent and open at their apices. He noted that a mature conidium lies behind the opening of the fertile hyphae and is pushed out by the younger conidium lying behind it. He described the conidia as cylindrical, truncate at both ends, one-celled, hyaline, 7-10 μ long and 3 μ wide, often occurring in very long chains of as many as 80 conidia. Amid these conidiophore elements, he observed perithecial primordia which later developed into brown, flask-shaped perithecia. He noted that the fungus did not produce chlamydospores, and placed it in a new genus, *Pyxidiophora* Bref., with *P. nyctalidis* Bref. as the type species. However, he did not name the conidial state. Lindau (1907) named the conidial state *Chalara brefeldi* Lindau and gave a description as follows: "Conidiophores clustered, brownish, paler at the apex. Conidia cylindrical, truncate at both ends, 10-15 μ long, 3-3.5 μ wide, often in branched chains, issuing from the interior of the phialides." Brefeld's description and illustrations leave little doubt that the fungus is a good *Chalara.* Unfortunately we have not been able to locate type or authentic material of this fungus. Nor does the description permit us to ascertain its affinities. So *Chalara brefeldi* must be treated as a *nomen dubium.*

5) Chalara cocos Pim *nom. nud.*

in Proc. R. Ir. Acad., ser. IV, 2: 27, 1884.

The name was based on a collection from Ireland. Pim (1884) stated that the fungus was doubtfully distinct from *Chalara fusidioides.* We have not been able to locate the original collection, and in the absence of adequate data we cannot dispose this fungus satisfactorily.

6) Chalara cyttariae Bomm. & Rouss.

in Bull. ad. r. Belg. cl. sci., p. 644, 1900; Sylloge Fung., 16: 1024, 1902.

A collection made on *Cyttaria darwini* in Tierra del Fuego was described as follows: "Caespitulis atris, effusis, oculo etiam armato inconspicuis; hyphopodio brunneo, septato, ramulis brevibus, fuscis, inflatis, 15-30 x 6, in basi peritheciorum Coniothyrii repentibus; hyphis fertilibus erectis, brevissimis, simplicibus, sursum attenuatis, hyalinis, basi dense congestis et tunc fuligineis, 19-45 x 1-5; conidiis catenulatis, cylindraceis, 12-15 x 2.5, hyalinis, intra tubulo formatis ex apice

A) *Hormiactis* sp. or *Anungitea* sp. ex type of *Chalara ampullula* var *minor* in PAD.
B) *Chalara heterospora* B. ex type.
C) *Chalara heterospora* ex F H 1612.

hypharum exsilientibus, deinde in articulos subellipsoideos secedentibus."

We cannot locate the type specimen. The description suggests its affinity with *Chalara,* but does not permit proper identification, so we must consider *C. cyttariae* a *nomen dubium.*

7) **Chalara drosodes** Höhn.
in Sber. Akad. Wiss. Wien, Abt. I, 127: 42, 1918.

Höhnel observed this fungus when examining a specimen (ex Krieger, F. Saxon. Exs. Nr. 1876) determined by Rehm as *Helotium drosodes* Rehm. Höhnel noted the presence of *Chalara* phialides on the apothecia, characteristically concluded that it was the imperfect state of the ascomycete, and gave it the name *Chalara drosodes* Höhn. He stated that the fungus was found on the outside of the excipulum and often on the short, thick stipe. The fertile hyphae were characterized as thin-walled, dark brown, stiff, erect or slightly bent, 60-110 μ long, 6 μ wide at the base, with 2-3 transverse septa, terminating in an obclavate open cell [phialide] 50-90 μ long, about 9 μ wide at its base, and 3-5 μ wide at the apex. The conidia were 3-4 μ long and 2 μ wide, and formed a chain.

Portions of Rehm's collection were reported to be available in W and WU, but are not traceable in these herbaria. Since we cannot ascertain the status of *C. drosodes,* we regard it as a *nomen dubium.*

8) **Chalara fusidioides** (Cda.) Rabenh. var. **longior** Sacc.
in Sylloge Fung., 19: 270, 1910.

Saccardo (1876) recorded what he identified as *Chalara fusidioides* Cda. (*sic*) from a collection of rotting *Quercus* wood in Italy. He noted that the 'hyphae' were dark and not hyaline as described by Corda, and that his fungus was very close to *Sporoschisma ampullula* and *S. montellica.* He made the following observations (1877) on the same fungus: "conidia cylindrica, utrinque truncata, 18-20 x 2, concatenata, hyalina. Hyphae phialiformes, 40 x 2.5-3 (in collo), fuscidulae. Conidia sec Corda 7-8 lga. et hyphae 21 lgae; an ex errore mensionis ? an species diversa ? Habitu tamen simillimus." Later (1886) he listed the fungus under *C. fusidioides* (Cda.) Rabenh. with the binomial *Chalara longior* Sacc., but the rank he intended for it was not clear until 1897 when he listed it as a subspecies of *C. fusidioides.* Oudemans' (1924) use of the name *C. longior* Sacc. at the species level appears to be a *lapsus calami.* Saccardo (1901) had reduced *C. longior* to synonymy with *C. fusidioides,* but subsequently again (1910) raised it to the rank of variety of *C. fusidioides.*

We have not been able to locate Saccardo's original collection and so cannot make a proper disposition of the fungus.

9) **Chalara gigas** Rostrup
in Dansk. bot. Ark., 2: 46, 1916; Sylloge Fung., 25: 785, 1931.

Rostrup erected this species for a collection on *Acer pseudoplatanus* in Denmark. The description was very brief: "Caespitulis minutis, subfuscis; hyphis fertilibus sparsis, paene aequicrassis, obscure furcatis, 220-235 x 10; conidiis cylindraceis, hyalinis, 24-35 x 5." His illustration of the fungus shows a dark conidiophore, at the apex of which cylindrical, unicellular, hyaline conidia occur in chains. Neither the illustration nor the description is of any help in deciding whether or not the conidia are formed enteroblastically in phialides. Rostrup's collections are believed to be in C, but a request for the type specimen elicited the information that it is not available there. *C. gigas* is therefore treated as a *nomen dubium.*

10) Chalara heterospora Sacc. (Figure 58B)
in Michelia, 1: 80, 1877; Sylloge Fung., 4: 334, 1886.

Saccardo described this species from a collection on rotting *Quercus* wood in Italy. His description is as follows: "Effusa, velutina, brunneo fuliginea; hyphis erectis, cylindraceis, 70-80 x 4-5, usque ad mediam longitudinem incrassatis, 4-5-septatis, fuligineis; conidiis ex apice hypharum exsilientibus, catenulatis, cylindraceis, 10-15 x 2-3, utrinque truncatis, nunc continuis, nunc 1-3-septatis, hyalinis."

DAOM 43615 contains a slide labelled as "*Chalara heterospora* Sacc., on cut end of wood, *Sporoschisma* (?) *insigne* + querc. cf. ic. 98 Selv. 76. 10 ex type collection of *Chalara heterospora* Sacc." No phialides or conidia of a *Chalara* like that described by Saccardo were seen in this slide, but a few phialides and phialoconidia of a *Sporoschisma*, and a few conidia of *Bispora* sp., were observed. Through the courtesy of Dr. Tomaselli, the type specimen was obtained from PAD. The few intact conidiophores that could be found (Figure 58B), were simple, cylindrical, with a slightly inflated base, brown, up to 3-septate, 64-66 μ long, with a smooth wall up to 0.5 μ thick; terminating in a phialide. Phialides ampulliform, 42-45 μ long, with a conical venter 12-13 μ long and 6-7.5 μ wide, and a cylindrical collarette up to 33 μ long and 3.5-4 μ wide at its apex. Transition from venter to collarette gradual. Phialoconidia (Figure 58B) were cylindrical with truncate ends, 0-1-septate, hyaline, 8-13 x 2-3 μ, smooth-walled, occurring singly or in chains. A few 3-septate, cylindrical conidia up to 24 μ long, with an obconical apical cell and a conical basal cell, were observed. Saccardo's illustration shows a few such conidia, but he gave their maximum length as 15 μ. The depauperate type specimen furnishes only fragmentary taxonomic information.

Höhnel's collection in FH contains a specimen, 1612 in folder 11126, that was assigned to *C. heterospora* by Höhnel. The label on the packet, in Höhnel's handwriting, reads: "Herb. Prof. Dr. Fr. v Höhn. *Chalara heterospora* Sacc., *Solidago*, Schonbichl. b. Tulln., 26.X.1904, Donau Auen, v. Höhn." This is apparently the specimen to which Höhnel (1904) referred as *Chalara heterospora* Sacc. occurring on stalks of *Solidago serotina*, and gave measurements of conidiophores and two-celled conidia as 75 x 6-9.5 μ and 16 x 2.5 μ respectively. He also commented on this specimen as follows: "Die Form nähert sich etwas der affinis, welche auch meist zweizellige Sporen zeigt." This specimen also is very depauperate. The colony appears superficial, effuse, composed of solitary, scattered or clustered conidiophores. The conidiophores (Figure 58C) are simple, cylindrical, 50-85 μ long, 5-7 μ wide at the base, 2-3-septate, the septa barely discernible; brown or slightly reddish brown; wall smooth and 0.5 μ thick; terminating in a phialide. The phialides are more or less conical, 38-64 μ long, composed of a venter 13-28 μ long and 6.5-9 μ wide, and a subcylindrical collarette 19-35 μ long and 2.5-4 μ wide at the apex. A single, cylindrical, unicellular, hyaline conidium 10 x 2 μ, with a rounded apex and a truncate base, was seen. These observations are insufficient to identify this fungus as *C. heterospora*. We have not seen any other specimen that could be matched with Saccardo's description of *C. heterospora*, which is here treated as a *nomen dubium*.

11) Chalara kriegeriana Bres. (Figure 59)
in Hedwigia, 33: 210, 1894; Sylloge Fung., 11: 616, 1895.

Bresadola described this fungus from a collection made by Krieger on leaves of *Syringa vulgaris* L. in East Germany. The description is as follows: "Late effusa, oculis vix conspicua; hyphis fertilibus erectis, haud fasciculatis, e basi attenuatis, septatis, ferrugineo-fuscis, apice albidis, 100-120 x 4 μ; conidiis catenulatis ex apice hypharum exsilientibus, cylindricis, utrinque truncatis, chlorino-hyalinis, 4-5 x 2 μ."

Bresadola's herbarium in S has a specimen #66 labelled: "*Ascochyta syringae* Bres. n. sp. au = *Phyllosticta syringae* West. mal observata? A *Chalara kriegeriana* Bres. n. sp. auf *Syringa vulgaris* L. in Prossen bei Schandau, Sachsen, 16/10/1892 W. Krieger" and bearing pencil sketches and some notes. The sketches show a long, septate, hyphal element (100-120 x 4 μ) [conidiophore?] pigmented below and hyaline above; 0-1-septate, ovoid conidia (8-10 x 3-3.5 μ); and non-septate, cylindrical conidia with truncate ends (4-5 x 2 μ); the notes read: "chlorino hyalinis cylindraceis apicibus obtusis-subtruncatis mox in articulos duos utrinque truncatis 4-5 dilabentibus."

The packet contains very brittle leaf material, on some of which oval to irregular leaf spots bearing minute, brown to black pycnidia are visible. Examination of scrape mounts from the spotted tissues revealed long, septate, cylindrical to subcylindrical conidiophores, dark brown over most of their length, becoming subhyaline to hyaline apically, with slightly roughened walls. Many of these conidiophores are geniculate toward the apex. The conidia are sparse, ovoid to fusiform, hyaline, subhyaline or pale brown, 0-1-septate, mostly with rough walls. The poor condition of the specimen and the sparse fungal elements make it difficult to determine the

59) *Chalara kriegeriana* ex type.

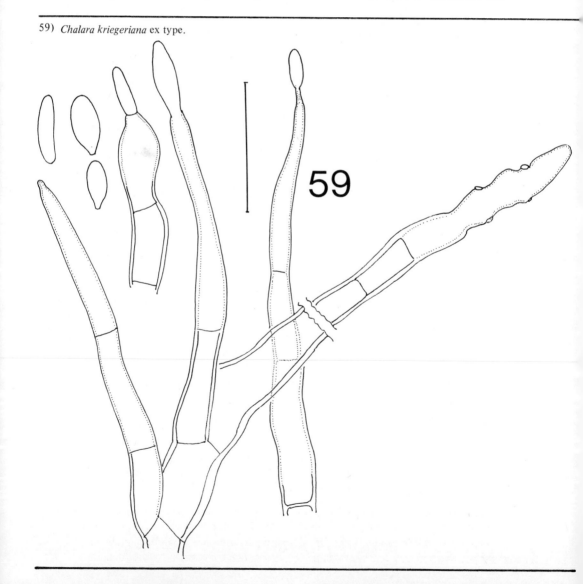

identities of the fungi present. The slides show a *Heterosporium*? sp., among other fungi. However, no *Chalara* was found, and *C. kriegeriana* is excluded from *Chalara.*

12) **Chalara longissima** Grove (Figure 60)
in J. Bot., Lond., 1885, p. 12.

Grove described the fungus from a collection made on rotting wood in Birmingham, England, as follows: "Hyphis fertilibus dense gregariis, erectis, rigidis, strictis, septatis, aequalibus, 150-170 x 4-5, infra fuscis, supra pallidioribus et saepe subinflatis, in catenulam conidiorum album longissimam flexuosam stipitem duplo super-autem evadentibus; conidiis fusoideis, irregularibus, utrinque subacutis, fere hyalinis, continuis, uni-quadri-guttulatis, 10-15 x 3.5-4." Grove included an illustration of the fungus, and considered it to have affinity with *Chalara longipes.*

The holotype specimen is IMI 17047. A prepared slide of the specimen shows a number of cylindrical, septate conidiophores (Figure 60A), 50-105 μ long, swollen at the base to a width of 3-4.5 μ, wall 0.5-1 μ thick; dark brown at the base, paler above, and terminating in a phialide. The phialide is cylindrical to subcylindrical, 24-42 μ long, 3-6 μ wide at the broadest part, composed of a long venter and an obconical collarette up to 9 μ long and 2-3 μ wide at the apex. The collarette wall is slightly verrucose. The phialoconidia are fusiform with truncate ends, unicellular, hyaline, 7-11 x 2-4 μ, with a smooth or sparsely echinulate wall; they originate successively at the base of the collarette and form short basipetal chains.

A similar fungus is present in IMI 32266 (Figure 60B) collected by S. J. Hughes on *Pteridium aquilinum* at Oxshott, Surrey, England, in 1948. In this specimen, the phialophores are 30-300 μ long, 3.5-4.5 μ wide at the base, and terminate in a subcylindrical phialide 21-50 μ long with a short, obconical collarette. The conidia are fusiform with truncate ends, unicellular, hyaline, 6.5-10 x 2.5-3 μ, with minutely echinulate walls; they occur in short basipetal chains. Percurrent proliferation of the phialophores is fairly common.

The long, subcylindrical venter with a relatively short, obconical collarette, and the absence of a really deep-seated conidiogenous locus, clearly exclude *C. longissima* from *Chalara.* Interestingly enough, it does not seem to fit into *Chloridium* or *Phialophora* either, as these genera are now circumscribed.

13) **Chalara maculicola** Moesz & Smarods apud Moesz
in Magy. bot. Lap., 33: 51, 1934.

This species, based on a collection made by Smarods on leaves of *Carpinus betulus* in Salaspils, Latvia, and noted to occur on leaf spots caused by *Gloeosporium robergei,* was described as follows: "Caespitulis maculicolis, amphigenis, minutis, sparsis, oculis inarmatis non visibilibus; conidiophoris ampulliformibus, basim inflatis, dilute brunneis, nonnunquam uniseptatis, supra in tubulum cylindraceum productis, hyalinis, totis 12-22 μ longis, basim usque ad 7.5 μ latis; conidiis ex apice exsilientibus, cylindricis, 3-7 μ x 2 μ, continuis, non catenulatis, indistincte guttulatis." The authors noted that this fungus was similar to *Chalara minima* Höhn., but differed from it in the length of the conidia.

A holotype specimen was not designated, but FH houses an exsiccatum labelled: "J. Smarods, Fungi Latvici Exsiccati. 896. *Chalara maculicola* Moesz et Smarods mit *Monostichella robergei* (Desm.) Höhn . . . Matrix: *Carpinus betulus* L., Prov. Vindzeme, Kr. Riga: Stopini, Baumschule Ch. Schoch, cult. 24.IX.32, leg. J. Smarods." This material bears a few phialides which are sessile, or borne on simple, up to 2-septate, stalks; ampulliform, brown, 19-23 μ long, with a slightly inflated venter 7.5-8.5 μ long and 3.5-4.5 μ wide, and a cylindrical to subcylindrical collarette 9-13 μ long and 2-2.5 μ wide. Transition from venter to collarette is very gradual; walls are smooth and 0.5 μ thick. Extremely few phialoconidia were observed. These are cylindrical with blunt ends, unicellular, hyaline, 11-13 x 1.5-4 μ. The description and the original illustration suggests that it is a good *Chalara,* but

60) ? *Chloridium* sp. A. ex IMI 17047; B. ex IMI 32266.

the paucity of the material necessitates relegation of *Chalara maculicola* to the status of a *nomen dubium*.

14) Chalara minima Höhn.
in Öst. bot. Z., 55 (p. 17—reprint), 1904; Sylloge Fung., 18: 629, 1906.

Höhnel described his new species from a collection on *Phialea sordida,* as follows: "Fruchthyphen durchscheinend braun, kleine zerstreute Rasen bildend, einzellig oder mit einer, selten zwei Querwänden an der Basis, 12-25 μ lang, oben 2 μ breit, nach abwärts keulig verbrotert 5-5½ μ dick, sehr dünnwandig, oben offen. Sporen aus dem Innern aus der Spitze der Fruchthyphen in kurzen Ketten heraustretend, hyalin, mit 1-2 öltröpfchen, einzellig, kurz, zylindrisch, 2-3 μ lang, 1½ μ breit."

Höhnel did not designate a holotype, but FH 1624, folder 11129, contains a packet labelled "Herb. Prof. Dr. Fr. v. Höhn. # 1624 *Chalara minima* v. H. n. sp., in margine et excipulo ascomatis *Phialeae sordidae,* Schönbichl, b. Tulln. 26.X.1904, Donau Auen,leg. v. Höhn." [scr. Höhn]. This is presumably an authentic specimen of *Chalara minima.* The packet contains a few twigs bearing apothecia of a discomycete. Careful examination revealed a few ampulliform phialides, so intimately mingled with the hyphal elements of the spothecia that it is difficult to observe the morphological features of the *Chalara* sp. The phialides appear to be sessile (?) smooth-walled, subhyaline to pale brown, 13-15 μ long, composed of a bulbous venter 6.5-8.5 μ long and 3.5-4.5 μ wide, and a cylindrical collarette 4.5-9 μ long and 2-2.5 μ wide. Phialoconidia are absent. The phialides resemble those of *C. fusidioides,* but in the absence of conidia we cannot draw a full comparison. If Höhnel's dimensions for the conidia are correct, then they are too small for *C. fusidioides.* There is no evidence that Höhnel distributed exsiccati of this fungus, or that other specimens exist. *Chalara minima* is here regarded as a *nomen dubium.*

FH houses another specimen collected and determined by Linder as *C. minima* v. Höhn. vel aff. This has been disposed under *C. microspora* (see page 125).

15) Chalara minuta Trail (*nom. nud.*)
in Scottish Naturalist 9: 41, 1887.

We have included what appears to be the original collection of *Chalara minuta* under *Chalara cylindrica* (see page 107). We are certain that *C. minuta* is identical with *C. cylindrica,* but since the former is a *nomen nudum* we have not listed it as a synonym of the latter.

16) Chalara montellica (Sacc.) Sacc.
in Michelia, 1: 80, 1877; Sylloge Fung., 4: 335, 1886.
≡ *Sporoschisma montellicum* Sacc.
in Nuovo G. Bot. ital., 7: 307, 1875; *ibid.,* 8: 191, 1876.

The fungus was found in decaying stems of *Melilotus officinalis* in Montello, Italy, in association with an ascomycete to which Saccardo gave the name *Rhaphidophora montellica* Sacc. [= *Ophiobolus* sp.]. The description was as follows: "hyphis prope perithecia crebriusculis, erectis e basi subincrassata cylindraceis, v. subfusoideis, 100-120 x 8, septatis, fuligineis, sursum attenuatis, pallidioribus; conidiis intra articulos supremos formatis et continuis ex apice exsilientibus cylindricis, 18-20 x 3, utrinque obtusis, 2-4 guttulatis, hyalinis." Saccardo illustrated the fungus in Figure 32 of his Fungi Italici, 1877. The type specimen is not available in any of the major Italian Herbaria. Saccardo's illustration is apparently of a good *Chalara,* but the description is inadequate to identify it with certainty. *C. montellica* is here considered a *nomen dubium.*

17) Chalara musae Sawada (*nom. nud.*)
in Spec. Bull. Coll. Agric. Nat. Taiwan Univ., 8: 193, 1959.

Sawada described this species, from a collection on dead petioles of *Musa cavendishi* in Nanton Prefecture in Formosa, as follows: "Without forming distinct lesion, fungus growing on leaf sheaths and petioles, spreading about 6 cm. broad, pubescent, dark brown; mycelia poor, subhyaline, 3-4 μ thick; conidiophores long, erected, simple or rarely branched, sometimes fasciculate, 3- to 10-septate, not constricted at the septum, dark brown to brown, 83-390 x 4.5-10 μ; ultimate branches 1-6, simple, verticillate or alternate, hyaline, obclavate, 13-26 x 3-4 μ; conidia formed endogenously at the apex of ultimate branches, cylindrical or elliptical, rounded at both ends, continuous, 4-7 x 1.5-2.5 μ." Sawada did not designate a holotype nor did he illustrate his fungus. In his publication, he mentioned that part of the type material had been given to the National Science Museum, Japan and that the original had been deposited in TAI. Dr. Tabuki could not find the specimens in the National Science Museum in Japan, and TAI did not respond to our request for a loan of the type specimen. Sawada noted: "The present fungus belongs to the genus *Chalara*, but the ultimate branches are singularly formed in the top and often 2-3-verticillate or sometimes alternate." These features exclude it from *Chalara*; it may possibly be a species of *Phialocephala, Sporendocladia,* or *Paecilomyces.*

18) Chalara mycoderma Bon.
in Handb. der Allgem. Mykologie, Stuttgart, p. 26, 1851.

The fungus was described by Bonorden (1851) from a collection made in Germany. Saccardo (1886) listed it as a synonym of *Geotrichum mycoderma* (Bon.) Sacc. and as a doubtful synonym of *Oospora lactis* (Fres.) Sacc. The name *C. mycoderma* appears to be of doubtful status. Carmichael (1957) has, however, treated it as a synonym of *Geotrichum candidum* Link ex Pers.

19) Chalara rivulorum Peyr.
in Nuovo G. bot. ital., 25: 442, 1918.

This fungus was collected on submerged, decorticated trunks of *Alnus viridis* and *Salix caprea* in Piedmont, Italy. Peyronel's description and illustration of the fungus show it to be a good *Chalara.* Peyronel did not designate a type and his specimens are not traceable in any of the major Italian Herbaria. The following account of the fungus is derived from Peyronel's description: Colony effuse, white, powdery. Phialophores gregarious, cylindrical, 1-7-septate, dark brown, 60-95 μ long and 6-7 μ wide, terminating in a phialide. Phialides lageniform, brown, 40-57 x 7-8.5 μ, composed of a subcylindrical venter and a long cylindrical collarette. Phialoconidia cylindrical, truncate at both ends, unicellular, hyaline, (3-)6-10(-13) x 2.5-3 μ, occurring in long chains.

In the absence of an authentic specimen we cannot determine this fungus.

20) Chalara strobilina Sacc.
in Nuovo G. bot. ital., 8: 185, 1876; Sylloge Fung., 4: 335, 1876.

Saccardo described this species, from a collection (in association with *Helotium strobilinum* [Fr.] Fuckel) made in Italy on rotting cones of *Abies excelsa,* as follows: "Effusa, velutina, brunnea; hyphis erectis obclavato-ampullaceis, 30-35 x 4 (in apice), 2-3-septatis, fuligineis; conidiis ex apice hypharum exsilientibus, catenulatis, cylindraceis, 4 x 1½, 2-guttulatis, hyalinis." The type specimen is not available in any of the major Italian herbaria consulted. Saccardo's illustration shows the fungus to be a good *Chalara,* but since the description is inadequate to identify it with certainty, *C. strobilina* is here considered as a *nomen dubium.*

21) Chalara terrestris Agnih. & Barua.
in Lloydia, 25: 173, 1962.

From the original description of the fungus collected on infected roots of *Camellia sinensis* in Assam, India, the following features are apparent: colonies sparse, pale brownish to silvery gray, velutinous, with scanty, intercellular, hyaline to subhyaline mycelium. Aerial mycelium sparse, fuscous. Phialophores cylindrical, 1-5-septate, dark brown and thick-walled at the base, pale brown and thin-walled toward the apex, 100-250 μ long, 2-3 μ wide at the base, terminating in a phialide. Phialides obclavate, composed of an ellipsoidal venter 5-8(-9) μ wide and a narrow, cylindrical collarette 2-3(-4) μ wide. Phialoconidia cylindrical with truncate ends, 1-septate, hyaline, 10-12 (-16) x 2-3(-4) μ, often catenate.

We have not been able to examine the type specimen, so cannot determine its affinities with any certainty.

22) Chalara variospora (Davids.?) Arn.
in Bull. Soc. mycol. Fr., 69: 278, 1953.

Chalara variospora appears to be a name of uncertain status. Arnaud (1953) published the name for two collections—GA #1591 and GA #2261—collection details for which were not given. The basionym was given as *Endoconidiophora variospora* Davids. (Mycologia 46: 303, 1944), the original description of which included that of its unnamed conidial state. Some nomenclatural problems are associated with this name. We have not been able to compare Arnaud's collections of the fungus with *Ceratocystis variospora* (Davids.) Moreau, which has been considered a synonym of *Ceratocystis fimbriata* by Webster and Butler (1967).

23) Sporoschisma connari Bat. & Peres apud Batista *et al.* in Publcões Inst. Micol. Recife, 298: 33, 1960.

Batista *et al.* (1960) described *Sporoschisma connari* occurring on leaves of *Connarus suberosus* in Brazil as follows: "Coloniae velutinae atro-brunneae in maculis rotundis, 2 mm. diam. Mycelium immersum, brunneum, ex hyphis in cellulis epidermicis evolutum. Conidiophori cylindraceo-tubuliformes continui, brunnei, 28-50 x 7-8.5 μ, in fasciculis densis. Conidiae endogenae, cylindraceae, 0-3-septatae, primo hyalinae dein brunneae, 13-17 x 5-6 μ. Hyphae capitatae non visae." Their illustration of the fungus shows a few phialides resembling those of a *Chalara* species or an *Ascoconidium* species with a few ellipsoidal or cylindrical conidia. Hughes (1966) excluded this taxon from *Sporoschisma* principally because it lacks capitate hyphae. Because of its overall resemblance to *Chalara*, attempts were made to study the type specimen. Prof. Carneiro informed us that the type specimen had been considerably depleted, and that only a slide prepared from it was available. The slide showed a few overlapping phialides and very few conidia. It is a good *Chalara,* but the meagre elements on the type specimen and the inadequacy of the single slide for characterization of the fungus necessitates that *S. connari* be considered a *nomen dubium.*

24) Thielaviopsis musarum (Mitch.) H. Riedl
in Sydowia, 15: 249-250, 1962.
≡ *Thielaviopsis paradoxa* (de Seynes) Höhn. var. *musarum* Mitchell (*nom. nud.*).
in J. Coun. scient. ind. Res. Austr., 10: 130, 1937.

Mitchell (1937) studied an isolate of *Thielaviopsis paradoxa* obtained from bananas affected by stem-end rot disease in Queensland, Australia. He reported the dimensions of phialoconidia and the dark thick-walled conidia as 9-23 x 3-6 μ and 10-32 x 5-13 μ respectively. He remarked that the isolate was not only slightly distinct from *T. paradoxa* in spore size, but also exhibited different host specificity in cross-inoculation tests involving banana, pineapple and sugarcane, and was slightly different in physiology. Mitchell considered these differences adequate to distinguish the isolate as a new variety of *Thielaviopsis paradoxa* and proposed the name *T. paradoxa* var. *musarum (nom. nud.)* for it.

Riedl (1962) observed a *Ceratocystis* on stem pieces of *Musa* in a grocery store in Vienna and considered it to be distinct from *C. paradoxa* (Dade) C. Moreau. He gave the name *Thielaviopsis musarum* (Mitchell) Riedl to its conidial state with the following description: "In speciminibus a me examinatis conidiophori terminales, elongatoconici, cellulis basalibus sterilibus 3-4 brunneis, 27.0-47.0 μ longis, 16.5-18 μ crassis, parte terminali aperto 170-200 μ longis, hyalinis vel basin versus brunnescenti; conidia minora primo subquadrangularia, hyalina, postea rotundata, fuliginea, rarius iam in conidiophoris dilute fuliginea, 9-12 μ longa, 6-7 μ crassa, demum usque ad 14 μ elongata, usque ad 9 μ dilatata; conidia minora origine non visa opace olivaceobrunnea, 11-14.5 μ longa, 7.3-9.2 μ crassa, guttulam oleosam unicam plerumque continentia." Although raising Mitchell's taxon to specific rank, Riedl had not even examined Mitchell's specimen.

We have not been able to locate the specimens of Mitchell or Riedl. There is little doubt that this is a good *Chalara*; the morphology and dimensions of the phialides, phialoconidia, and the thick-walled, dark conidia, fall well within the broad limits of *Chalara paradoxa*. But Riedl's conclusion that the perfect state is distinct from *Ceratocystis paradoxa* necessitates a reappraisal of the fungus before a definite name could be applied to the conidial state.

25) Thielaviopsis neo-caledoniae Dadant (*nom. nud.*)
in Revue gén. Bot., 57: 168, 1950.

Dadant (1950) observed a pathogenic fungus on *Coffea robusta* in New Caledonia and suggested the name *Thielaviopsis neo-caledoniae* "if it should prove to be a new species." He observed only the phialidic state, each of the phialides bearing a chain of cylindrical, unicellular, hyaline, smooth-walled conidia. On maize agar, the dimensions of phialides and phialoconidia were reported to be 55.5-81.4 x 4.6-6.2 μ and 3.7-11.1 x 1.9-2.8 μ respectively. These characters suggest a possible affinity with *Chalara*. We have been unable to locate the specimen. This is a name of doubtful nomenclatural and taxonomic standing.

26) Thielaviopsis podocarpi Petri
in Nuovo G. bot. ital., 10: 584, 1903.

Petri described this fungus from tuberculate roots of various species of *Podocarpus* as follows: "Hyphae steriles, repentes, subhyalinae, septatae, ramosae; fertiles erectae, breves, furcatae, septatae. Macroconidia catenulata, globosa, cuboidea, fusca; microconidia cylindracea, hyalina, utrinque truncata, 2 guttulata, ex hyphis fusoideis, septatis simplicibus vel furcatis, olivaceis, supra pallidioribus (μ = 160-180 = 4-5) generata. Macroconidia μ = 8.5 = 5.2. Microconidia μ = 2-2.5 = 8.5-9." Petri published an illustration of the fungus which shows two kinds of conidia. One kind, referred to as microconidia by Petri, are cylindrical with truncate ends, and hyaline, and originate from the apex of subulate conidiophores, in chains; the second kind occurs as rectilinear series of 4-5 dark coloured and thick-walled, cuboidal conidia on short conidiophores with a sympodial branching pattern. We have been unable to locate the type specimen in any of the major Italian Herbaria. Petri's inclusion of the fungus in *Thielaviopsis* suggests that the cylindrical conidia may be formed in phialides; the morphology of the fungus and dimensions of the conidia as given by Petri agree well with those of *Chalara elegans*. In the absence of the type specimen, we cannot place this fungus with any certainty. *T. podocarpi* is here treated as a *nomen dubium*.

Appendix
Calycellina carolinensis sp. nov. and *Hyaloscypha cladii* sp. nov.

On page 55, we mentioned the consistent association of discomycetes with two species of *Chaetochalara*. Apothecia of two discomycetes were found to be intimately associated with, and may indeed have arisen from the same mycelium as the *Chaetochalara* species involved. Since these very reduced discomycetes have few good characters left in the ascocarp, a precise knowledge of their conidial states, where such exist, would be of considerable taxonomic value. Although we are not in a position to substantiate genetic relationships between the discomycetes and *Chaetochalara,* the possibility of such a relationship cannot be excluded in view of the manifest association in two geographically disparate collections. Accordingly, we felt the need to establish the identity of these discomycetes. We are grateful to Dr. R. W. G. Dennis, Royal Botanical Gardens, Kew, Surrey, England, who has kindly examined the specimens and has advised us that the discomycetes represent undescribed species of *Calycellina* and *Hyaloscypha.*

Calycellina carolinensis sp. nov. (Figure 61A)
Apothecia vulgo hypophylla nonnunquam amphigena, superficialia, sessilia, flava vel brunnea cum annulo fuscato basilium cellularum, pallidis, septatis, laevibus, obtusis, marginalibus, 24-40 x 3-4 μ, pilis tecta discus planus, pallidus, 170-290 μ diam. Asci clavati, hyalini, 8-spori, 45-70 x 6.5-8 μ. Ascosporae anguste fusiformes, leniter curvatae, 1-septatae, hyalinae, pariete laevi, 12-17 x 2.5-3.5μ biseriatim vel irregulatim dispositae. Paraphyses filiformes, hyalinae, septatae, pariete laevi, usque ad 70 μ long. et 2-2.5 μ lat.

Apothecia amphigenous, mostly hypophyllous, superficial, sessile on a small base, pale yellow to brown with a dark ring of basal cells, covered with pale, septate, smooth, blunt marginal hairs, 25-40 x 3-4 μ; disc flat, pale yellow, 170-290 μ diam. Asci clavate, 45-70 x 6.5-8 μ, hyaline, smooth-walled, octosporous. Ascospores narrowly fusiform, slightly curved, 1-septate, hyaline, 12-17 x 2.5-3.5 μ, biseriate or irregularly arranged. Paraphyses filiform, septate, hyaline, up to 70 μ long and 2-2.5 μ wide.

Habitat: On leaves of *Knightia excelsa* and *Myrica cerifera,* in association with *Chaetochalara aspera.*

Specimens examined: DAOM 139268 [*Holotype*] , on leaf of *Myrica cerifera,* Research Triangle Pk., North Carolina, U.S.A., 11.IX.1972, C. S. Hodges; PDD 32877, on leaf of *Knightia excelsa,* Kauaeranga Valley, Thames Co., N.Z., 21.I.1974, B. Kendrick (KNZ 213).

Known distribution: New Zealand, U.S.A.

Hyaloscypha cladii sp. nov. (Figure 61B)
Apothecia dissita, sessilia vel breviter stipitata, poculiformia, brunnea, ad marginem basaliter septatos, pallide brunneos, subulatos pilos, 40-75 x 2.5-3.5 μ ferentia; discus concavus, pallidus, ca. 170 μ diam. Asci clavati, hyalini, 8-spori, 44-60 x 10-12 μ. Ascosporae ellipsoideae, 1-septatae, hyalinae, guttulatae, pariete laevi, 12-16 x 4-5 μ, biseriatim vel irregulatim dispositae. Paraphyses filiformes, hyalinae, sparsim septatae, pariete laevi, 40-60 x 1.5-2 μ.

Apothecia scattered, sessile to shortly stalked, cup-shaped, brown, bearing, near the margin, basally septate, pale brown, subulate hairs, 40-75 x 2.5-3.5 μ; disc concave, pallid, about 170 μ diam. Asci clavate, hyaline, octosporous, 44-60 x 10-12 μ. Ascospores ellipsoidal, 1-septate, hyaline, guttulate, smooth-walled,

12-16 x 4-5 μ, biseriate or irregularly arranged. Paraphyses filiform, hyaline, sparsely septate, smooth-walled, 40-60 x 1.5-2 μ.

Habitat: On *Cladium mariscus,* in association with *Chaetochalara cladii.*

Specimen examined: IMI 89626(b) [*Holotype*]   Sugar Hill, Wareham, Dorset, 26.IV.1961.

Known distribution: U.K.

61A) *Calycellina carolinensis.* Vertical section of an apothecium, seta, marginal hairs, paraphysis, ascus and mature ascospores ex type in DAOM 139268.

61B) *Hyaloscypha cladii.* Vertical section of an apothecium, setae, marginal hairs, paraphysis, ascus, and mature ascospores (illustration prepared by Dr. K. A. Pirozynski, DAOM, Ottawa, and reproduced here with his kind permission).

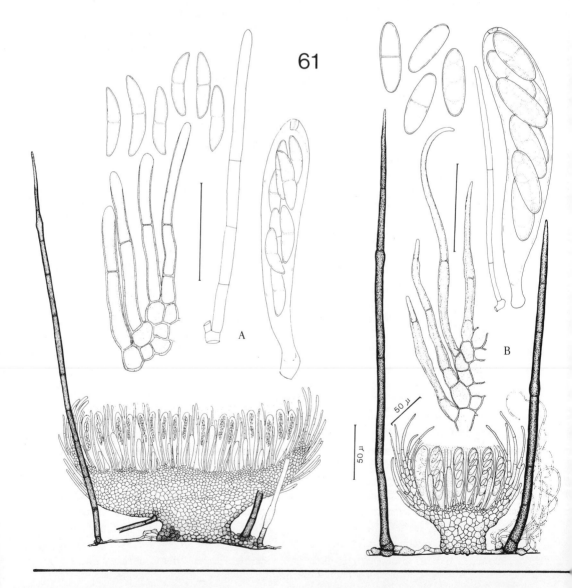

References

Ainsworth, G. C. 1971. Ainsworth and Bisby's Dictionary of the Fungi. 6th Edition. Commonwealth Mycological Institute, Kew, England.

Andrus, C. F. 1933. Morphology and reproduction in *Ceratostomella fimbriata*. J. agric. Res., 46: 1059-1078.

Arnaud, G. 1953. Mycologie concréte: Genera II (Suite et fin.) Bull. Soc. mycol. Fr., 69: 265-306.

Arnaud, G. and Barthelet, J. 1936. Le résumé ou pourriture noir châtagnes *Sclerotinia pseudotuberosa* et *Rhacodiella castaneae*. Annls Épiphyt., N.S., 1: 121-146.

Bakshi, B. K. 1950. Fungi associated with ambrosia beetles in Great Britain. Trans. Br. mycol. Soc., 33: 111-120.

Barron, G. L. 1968. The genera of Hyphomycetes from soil. The Williams and Wilkins Company, Baltimore. 366 pp.

Batista, A. C. and Vital, F. A. 1956. *Hughesiella*—novo e curioso genero de fungos dematiaceae. Anais Soc. Biol. Pernamb., 14: 141-144.

Berkeley, M. J. and Broome, C. E. 1850. Notices of British Fungi. Ann. Mag. nat. Hist. ser. 2, 5: 455-466.

Berkeley, M. J. and Broome, C. E. 1854. Notices of British Fungi. Ann. Mag. nat. Hist., ser. 2, 13: 458-469.

Bliss, D. E. 1941. A new species of *Ceratostomella* on date palm. Mycologia, 33: 468-482.

Bonorden, H. F. 1851. Handbuch der Allgemeinen Mykologie. Stuttgart. 336 pp.

Brefeld, O. 1891. Untersuchungen aus dem Gesammtgebiet der Mykologie. Heft X. Ascomyceten II. Münster. 212 pp.

Brierley, W. B. 1925. The endoconidia of *Thielavia basicola* Zopf. Ann. Bot., 29: 483-493.

Bubák, F. 1916. Pilze von verschiedenen Standorten. Annls. mycol., 14: 341-352.

Carmichael, J. W. 1957. *Geotrichum candidum*. Mycologia, 49: 820-830.

Clements, F. E. and Shear, C. L. 1931. The genera of fungi. H. W. Wilson Company, New York. 496 pp.

Cole, G. T. and Kendrick, W. B. 1968. A thin culture chamber for time-lapse photomicrography of fungi at high magnifications. Mycologia, 60: 340-344.

Cole, G. T. and Kendrick, W. B. 1969. Conidium ontogeny in hyphomycetes. The phialides of *Phialophora, Penicillium* and *Ceratocystis*. Can. J. Bot., 47: 779-789.

Cole, G. T. and Kendrick, B. 1973. Taxonomic studies of *Phialophora*. Mycologia 65: 661-688.

Cole, G. T., Nag Raj, T. R. and Kendrick, W. B. 1969. A simple technique for time-lapse photomicrography of microfungi in plate culture. Mycologia, 61:

726-730.

Cooke, M. C. 1885. New British Fungi. Grevillea, 14: 1-7.

Corda, A. C. J. 1838. Icones fungorum hucusque cognitorum. 2. Prague.

Corda, A. C. J. 1842. Icones fungorum hucusque cognitorum. 5. Prague.

Dadant, R. 1950. Sur une maladie du *Coffea robusta* en Nouvelle Caledonie. Revue gén. Bot., 57: 168-176.

Dade, H. A. 1928. *Ceratostomella paradoxa*, the perfect state of *Thielaviopsis paradoxa* (de Seynes) v. Höhn. Trans. Br. mycol. Soc., 13: 184-194.

Davidson, R. W. 1944. Two American hardwood species of *Endoconidiophora* described as new. Mycologia, 36: 300-306.

Delacroix, M. G. 1893. Éspèces nouvelles observées au Laboratoire de Pathologie végétale. Bull. Soc. mycol. Fr., 9: 184-188.

De Seynes, J. 1886. Recherches pour servir à l'Histoire naturelle des végétaux inférieurs. III. 1re Partie, Paris. 85 pp.

El Ani, A. S. 1958. The genetics of *Ceratostomella radicicola* and the phylogenetic relationship between *Chalaropsis* and *Chalara*. Am. J. Bot., 45: 228-232.

Ferdinandsen, C. C. F. and Winge, O. 1910. Fungi from Prof. Warming expedition to Venezuela and the West Indies. Bot. Tidsskr., 30: 208-222.

Ferraris, T. 1910. Flora Italica Cryptogama. Pars. I: Fungi, Hyphales, Tuberculariaceae-Stilbaceae. Fasc. No. 6, 979 pp.

Funk, A. 1966a. The type species of *Ascoconidium*. Can. J. Bot., 44: 39-41.

Funk, A. 1966b. *Ascoconidium tsugae* n. sp. associated with bark diseases of Western Hemlock in British Columbia. Can. J. Bot., 44: 219-222.

Funk, A. 1975. *Sageria*, a new genus of Helotiales. Can. J. Bot. 53: 1196-1199.

Groves, J. W. and Elliott, M. E. 1969. Notes on *Ciboria rufo-fusca* and *C. alni*. Friesia 9: 29-36.

Hammond, J. B. 1935. The morphology, physiology and the mode of parasitism of a species of *Chalaropsis* infecting nursery walnut trees. J. Pomol., 13: 81-107.

Hennebert, G. L. 1967. *Chalaropsis punctulata*, a new hyphomycete. Antonie van Leeuwenhoek, 33: 333-340.

Hennebert, G. L. 1968. *Echinobotryum, Wardomyces* and *Mammaria*. Trans. Br. mycol. Soc., 51: 749-762.

Hennebert, G. L. 1971. Pleomorphism in Fungi Imperfecti. pp. 202-223 in Taxonomy of Fungi Imperfecti. (Ed.) Bryce Kendrick. University of Toronto Press, Toronto.

Henry, B. W. 1944. *Chalara quercina* n. sp., the cause of oak wilt. Phytopathology, 34: 631-635.

Höhnel, F. von. 1902. Fragmente zur Mykologie. I Mitteilung. Sber. Akad. Wiss. Wien, 111: 987-1056.

Höhnel, F. von. 1904. Zur Kenntnis einiger Fadenpilze. Hedwigia, 43: 295-299.

Höhnel, F. von. 1909. Fragmente zur Mykologie. VI Mitteilung. Sber. Akad. Wiss. Wien, 118: 1-178.

Höhnel, F. von. 1910. Fragmente zur Mykologie. XI Mitteilung. Sber. Akad. Wiss. Wien, 119: 1-63.

Höhnel, F. von. 1925. Neue Fungi Imperfecti. Mitt. bot. Inst. tech. Hochsch. Wien, 2: 33-64.

Hughes, S. J. 1949. Studies on microfungi. II. The genus *Sporoschisma* Berkeley and Broome and a redescription of *Helminthosporium rousselianum* Montagne. Mycol. Pap. 31, 33 pp.

Hughes, S. J. 1953. Conidiophores, conidia and classification. Can. J. Bot., 39: 577-659.

Hughes, S. J. 1966. New Zealand Fungi. 6. *Sporoschisma* Berk. and Br. N.Z. Jl. Bot. 4: 77-85.

Hughes, S. J. and Nag Raj, T. R. 1973. New Zealand Fungi. 20. *Fusichalara* gen. nov. N.Z. Jl. Bot. 11: 661-671.

Hunt, J. 1956. Taxonomy of the genus *Ceratocystis*. Lloydia, 19: 1-59.

Illman, W. I. 1964. *Endosporostilbe,* an apparently superfluous generic name. Mycologia, 56: 920-921.

Illman, W. I. and Hamly, D. H. 1948. A report on Ridgway colour standards. Science, 107: 626-628.

Kendrick, W. B. (ed.). 1971. Taxonomy of Fungi Imperfecti. University of Toronto Press, Toronto. 309 pp.

Kendrick, W. B. and Carmichael, J. W. 1973. Hyphomycetes. pp. 323-509 in The Fungi, Vol. IV A. (Eds.) Ainsworth, G. C., Sparrow, F. K. and Sussman, A. S. Academic Press, New York.

Lindau, G. 1907. Kryptogamenflora von Deutschland, Österreich und der Schweiz. Fungi Imperfecti, Hyphomycetes, 8 Abt., Leipzig. 851 pp.

Malcolm, W., Noble, M. and Gray, E. 1954. *Gloeotinia*—a new genus of Sclerotiniaceae. Trans. Br. mycol. Soc., 37: 29-32.

Mason, E. W. 1937. Annotated accounts of fungi received at the Imperial Mycological Institute. List II, fasc. III (General Part). Mycol. Pap. 3, 31 pp.

Mason, E. W. 1941. Annotated accounts of fungi received at the Imperial Mycological Institute. List II, fasc. III (Special Part). Mycol. Pap. 4, 44 pp.

Massee, G. 1884. Description and life history of a new fungus, *Milowia nivea*. Jl. R. microsc. Soc., 4: 841-845.

Massee, G. 1893. British Fungus Flora. Vol. III, London. 512 pp.

McAlpine, D. 1902. Fungus diseases of stone fruit trees in Australia and their treatment. Melbourne. 165 pp.

McCormick, F. A. 1925. Perithecia of *Thielavia basicola* Zopf in culture. Bull. Conn. agric. Exp. Stn., 269.

Meyer, J. 1959. Moisissures du sol et litières de la region de Yangambi (Congo Belge). Publ. Inst. natn. Étude agron. Congo belge Ser. Scient. 75 pp.

Mitchell, R. S. 1937. Stem end rot of bananas with special reference to the physiological relationships of *Thielaviopsis paradoxa* (de Seynes) von Höhnel. J. Coun. scient. ind. Res. Aust., 10: 123-130.

Münch, E. 1907. Die Blaufaule des Nadelholzes. Naturw. A. Forst–Z. Landw., 5: 531-573.

Nag Raj, T. R. and Hughes, S. J. 1974. New Zealand Fungi 21. *Chalara* (Corda) Rabenhorst. N.Z. Jl. Bot. 12: 115-129.

Nag Raj, T. R. and Kendrick, W. B. 1971. On the identity of three species of *Cylindrosporium* described by Preuss. Can. J. Bot., 49: 2119-2122.

Orpurt, P. A. and Curtis, J. T. 1957. Soil microfungi in relation to the prairie continuum in Wisconsin. Ecology, 38: 628-637.

Oudemans, C. A. J. A. 1924. Enumeratio systematica Fungorum. Nijhoff, Den Haag. Vol 5. 998 pp.

Patouillard, N. and Lagerheim, G. 1891. Champignons de l'Equateur. Bull. Soc. mycol. Fr., 7: 158.

Petri, L. 1903. Di una nuova species di *Thielaviopsis* Went. Nuovo G. bot. ital., 10: 582-584.

Peyronel, B. 1916. Una nova malattia del lupino prodotta da *Chalaropsis thielavioides* Peyr. nov. gen. et nova sp. Staz. sper. agr. ital., 49: 583-595.

Pim, G. 1884. Recent additions to the fungi of counties Dublin and Wicklow [1883]. Proc. R. Ir. Acad. ser. IV, 2: 25-28.

Pirozynski, K. A. and Hodges, C. S. Jr. 1973. New Hyphomycetes from South

Carolina. Can. J. Bot., 51: 157-173.

Pirozynski, K. A. and Morgan-Jones, G. 1968. Notes on microfungi III. Trans. Br. mycol. Soc., 51: 185-206.

Plowright, C. B. 1884. A monograph of British *Hypomyces.* Grevillea, 2: 1-8.

Prillieux, E. E. and Delacroix, M. G. 1891. *Endoconidium temulentum* nov. gen., nov. sp., Prillieux et Delacroix, champignon donnant au seigle des propriétés vénéneuses. Bul. Soc. mycol. Fr., 7: 116-117.

Prillieux, E. E. and Delacroix, M. G. 1892. *Phialea temulenta* nov. sp. Prillieux et Delacroix état ascospore d'*Endoconidium temulentum*, champignon donnant au seigle des propriétés vénéneuses. Bull. Soc. mycol. Fr., 8:22-23.

Rabenhorst, L. 1844. Kryptogamenflora. I Bd. Pilze. Leipzig. 614 pp.

Riedl, H. 1962. *Ceratocystis musarum* sp. n., die Hauptfruchtform der *Thielaviopsis*—Art von Bananenstielen. Sydowia, 15: 247-251.

Roldan, E. F. 1962. Species of *Ceratocystis (Ceratostomella)* causing stain in Rattan. Philipp. J. Sci. 91 (4): 415-423.

Saccardo, P. A. 1875. Fungi Veneti novi vel critici. Ser. IV. Atti Accad. scient. veneto-trent.-istriana, 4: 101-141.

Saccardo, P. A. 1876. Fungi Veneti novi vel critici. Serv. V. Nuovo G. bot. ital., 8: 161-211.

Saccardo, P. A. 1877. Fungi italici autographice delineati. Michelia, 1: 1-116.

Saccardo, P. A. 1880. Conspectus generum fungorum Italiae inferiorum, nempe ad sphaeropsideas, melanconieas et hyphomyceteas pertinentium, systemate sporologico dispositorum. Michelia 2: 1-38.

Saccardo, P. A. 1886. Sylloge Fungorum omnium hucusque cognitorum. Pavia. Vol. 4. 807 pp.

Saccardo, P. A. 1897. Sylloge Fungorum omnium hucusque cognitorum. Pavia. Vol. 12. 1053 pp.

Saccardo, P. A. 1901. Sylloge Fungorum omnium hucusque cognitorum. Pavia. Vol. 15. 455 pp.

Saccardo, P. A. 1910. Sylloge Fungorum omnium hucusque cognitorum. Pavia. Vol. 19. 1158 pp.

Saccardo, P. A. 1911. Notae mycologicae. Annls mycol., 9: 249-257.

Savile, D. B. O. 1968. Possible interrelationships between fungal groups. pp. 649-675 in The Fungi, Vol III. (Eds.) Ainsworth, G. C. and Sussman, A. S. Academic Press, New York.

Seaver, F. J. 1942. Photographs and descriptions of cup fungi. 37. *Pezicula purpurascens.* Mycologia, 34: 412-415.

Stevens, F. L. 1925. Hawaiian Fungi. Bull. Bernice P. Bishop Mus., 19: 1-189.

Subramanian, C. V. 1958. Hyphomycetes V. J. Indian bot. Soc., 37: 47-64.

Sugiyama, J. 1968. Mycoflora in core samples from stratigraphic drillings in middle Japan. III. The taxonomic status of the genus *Chalaropsis* Peyronel (Hyphomycetes). J. Fac. Sci. Tokyo Univ., sec. 3, 1: 29-48.

Sutton, B. C. 1973. Hyphomycetes from Manitoba and Saskatchewan, Canada Mycol. Pap. 132. 143 pp.

Sutton, B. C. and Pirozynski, K. A. 1965. Notes on microfungi. II. Trans. Br. mycol. Soc., 48: 349-366.

Trail, J. W. H. 1887. Report for 1886 on the fungi of East of Scotland. Scott. Nat., 9: 39-42.

Tsao, P. H. and Bricker, J. L. 1970. Acropetal development in the 'chain' formation of chlamydospores of *Thielaviopsis basicola.* Mycologia, 62: 960-966.

Tulasne, L. R. and Tulasne, C. 1865. Selecta fungorum carpologia. Paris. Vol. 3. 221 pp.

Unger, D. F. 1847. Botanische Beobachtungen. Bot. Ztg., 15: 249-257.

Vuillemin, P. 1911. Les Aleuriosporés. Bull. Séanc. Soc. Sci. Nancy, 12 (3): 151-175.

Wakefield, E. M. and Bisby, G. R. 1941. List of Hyphomycetes recorded for Britain. Trans. Br. mycol. Soc., 25: 49-126.

Webster, R. K. and Butler, E. E. 1967. A morphological and biological concept of the species *Ceratocystis fimbriata*. Can. J. Bot., 45: 1457-1468.

Went, F. A. 1893. Die Ananasziekte van het suikerriet. Meded. Proefstn, Suik-Riet W. Java, V.

Whetzel, H. H. 1937. *Septotinia,* a new genus of Ciboriaceae. Mycologia, 29: 128-146.

Zopf, W. 1876. *Thielavia basicola* Zopf genus novum perisporiacearum. Sber. bot. Prov. Brandenberg, 18: 101-105.

Index of Genera and Species

Accepted names are printed in roman type and synonyms in *italics*. Nomina nuda, illegitimate names, doubtful and excluded taxa are indicated by an asterisk (∗) and *italics*. Page references in **bold-face** type indicate **descriptions**, those in *italics* indicate *illustrations*.

Substrate Index

This index is partly based on information derived from Review of Applied Mycology: Plant-Host index to Vols. 1-40, 1922-1961, and subsequent issues of RAM published up to 1968. Many of the records not verified by us are indicated by an asterisk (*). The name of the fungus in *italics* follows that of the substrate in roman type.

Daucus carota
Ceratocystis fimbriata
Chalara elegans
Chalara thielavioides

Dracophyllum traversii
Chalara curvata

Dracophyllum sp.
Chalara rhynchophiala

Dysoxylum spectabilis
Sporoschisma mirabile

Elaeis sp.
Ceratocystis paradoxa
Chalara elegans

Elaeis guineensis
Chalara paradoxa

*Eleocharis tuberosa
Ceratocystis adiposa
Ceratocystis paradoxa

Epilobium hirsutum
Sporoschisma mirabile

*Erythrina edulis
Ceratocystis paradoxa

Eucalyptus sp.
Chalara cylindrica

*Euphorbia pulcherrima
Chalara elegans

Experimental hosts of *Chalara elegans*
Cucurbita moschata
Mangifera indica
Musa sp.
Phoenix sp.
Rhapis sp.
Saccharum spontaneum

Fagus sp.
Chalara cylindrosperma
Chalara quercina
Chalara spiralis

Fagus sylvatica
Chalara affinis
Chalara cylindrosperma
Chalara ovoidea
Sporoschisma juvenile
Sporoschisma mirabile

*Fragaria sp.
Ceratocystis adiposa
Chalara elegans

Fragaria vesca
Chalara fusidioides

Fraxinus sp.
Chalara breviclavata
Chalara quercina

Fraxinus excelsior
Sporoschisma juvenile
Sporoschisma mirabile

Fungi—
On Hydnum sp.
Chalara microspora
On Hydnum compactum
Chalara fungorum
On old perithecia of Mycosphaerella spp.
occurring on Brya purpurascens, and
Tofieldia pusilla:
Chalara fusidioides

Geostachys rupestris
Chalara rostrata

*Gerbera jamesonii
Chalara elegans

Ginkgo biloba
Chalara ginkgonis

Gleditschia triacanthos
Chalara ampullula
Chalara cylindrosperma

Gloxinia sp.
Chalara elegans

*Gliricidia sepium
Ceratocystis fimbriata

*Gossypium sp.
Chalara elegans

Hedera helix
Sporoschisma juvenile

Heracleum giganteum
Sporoschisma mirabile

Herbaceous stems, undetermined
Chalara pteridina
Chalara urceolata

Hevea brasiliensis
Chalara fimbriata

Hoheria angustifolia
Chalara hughesii

Ilex aquifolium
Chaetochalara bulbosa

Ilex denticulata
Chalara cylindrosperma

Ilex europaeus
Chalara aurea

Ipomoea batatas
Ceratocystis fimbriata
Chalara paradoxa

Juglans sp.
Chalara thielavioides

Knightia excelsa
Calycellina carolinensis
Chaetochalara aspera
Chalara angionacea
Chalara gracilis
Chalara sessilis
Chalara tubifera
Sporoschisma mirabile

Podocarpus spicatus
Chalara bicolor

Podocarpus totara
Chalara cylindrosperma

*Poncirus trifoliatus
Chalara elegans

*Populus sp.
Ceratocystis fimbriata
Ceratocystis moniliformis

Primula sp.; *P. obconica;
*P. polyantha
Chalara elegans

*Prunus americana
Ceratocystis fimbriata

*Prunus cerasus
Chalara quercina

*Prunus persica
Chalara thielavioides

Pseudotsuga [taxifolia] menziesii
Ceratocystis coerulescens

Pteridium aquilinum
Chalara pteridina

*Pueraria javanica
Ceratocystis paradoxa

*Pyrus sp.
Chalara quercina

Pyrus malus
Ceratocystis fagacearum
Chalara quercina
Sporoschisma saccardoi

Quercus sp.
Ceratocystis fagacearum
Ceratocystis fimbriata
Ceratocystis moniliformis
Chalara affinis
Chalara aurea
Chalara brevispora
Chalara fusidioides
Sporoschisma juvenile
Sporoschisma mirabile

Quercus densiflora
Chaetochalara setosa

Quercus incana
Chalara emodensis

*Rhapis sp.
Chalara paradoxa

Rhipogonum scandens
Chalara dictyoseptata
Sporoschisma mirabile

Rhopalostylis sapida
Chalara urceolata
Sporoschisma mirabile

*Ricinus communis
Chalara elegans

Robinia pseudacacea
Chalara ampullula

*Rosa sp.
Chalara thielavioides

*Roystonea regia
Ceratocystis paradoxa

Rubus sp.
Chalara kendrickii
Chalara rubi

Rubus fruticosus
Chalara bohemica

Rumex sp.
Chalara urceolata

Saccharum officinarum
Ceratocystis adiposa
**Ceratocystis paradoxa*
**Sporoschisma nigroseptatum*

Salix sp.
Bloxamia truncata
Chalara ampullula
Sporoschisma mirabile
Sporoschisma saccardoi

Selaginella rupestris; S. arenicola
ssp. acanthonota
Chalara selaginellae

Senecio jacobina
Chalara pteridina

Soil, isolated from:
Chalara brevispora
Chalara elegans
Chalara ellisii

*Sagittaria sagittifolia
Ceratocystis paradoxa

Sassafras sp.
Chalara quercina

*Scabiosa sp.
Chalara elegans

*Scindapsus aureus
Chalara elegans

*Solanum esculentum
Chalara elegans

*Sorbus aucuparia
Sporoschisma juvenile
Sporoschisma mirabile

*Synangium auritum
Ceratocystis fimbriata

*Syringa vulgaris
Chalara elegans